Genes of Eden

Genes of Eden

Epigenetics, Transgenerational Sin, and Addiction

LUMAN R. WING
Foreword by Heather Falk

WIPF & STOCK · Eugene, Oregon

GENES OF EDEN
Epigenetics, Transgenerational Sin, and Addiction

Copyright © 2025 Luman R. Wing. All rights reserved. Except for brief quotations in critical publications or reviews, no part of this book may be reproduced in any manner without prior written permission from the publisher. Write: Permissions, Wipf and Stock Publishers, 199 W. 8th Ave., Suite 3, Eugene, OR 97401.

Wipf & Stock
An Imprint of Wipf and Stock Publishers
199 W. 8th Ave., Suite 3
Eugene, OR 97401

www.wipfandstock.com

PAPERBACK ISBN: 978-1-6667-5194-9
HARDCOVER ISBN: 978-1-6667-5195-6
EBOOK ISBN: 978-1-6667-5196-3

Cataloguing-in-Publication data:

Names: Wing, Luman R., author. | Falk, Heather, foreword author.

Title: Genes of eden : epigenetics, transgenerational sin, and addiction / Luman R. Wing ; foreword by Heather Falk.

Description: Eugene, OR : Wipf & Stock, 2025 | Includes bibliographical references.

Identifiers: ISBN 978-1-6667-5194-9 (paperback) | ISBN 978-1-6667-5195-6 (hardcover) | ISBN 978-1-6667-5196-3 (ebook)

Subjects: LCSH: Epigenetics. | Bioethics. | Christian ethics. | Substance abuse—Religious aspects—Christianity. | Drug addiction—Religious aspects—Christianity.

Classification: HV5186 .W56 2025 (paperback) | HV5186 (ebook)

VERSION NUMBER 09/05/25

Scripture taken from the New King James Version®. Copyright © 1982 by Thomas Nelson. Used by permission. All rights reserved.

For my priceless wife, Pamela, my sons Daniel and Aaron and my daughters, Laura and Heather, my son-in-law Chris, and my three grandsons, Lewis, Elliot and Micah.

"Understanding is the reward of faith. Therefore, seek not to understand that you may believe, but believe that you may understand."

— St. Augustine of Hippo

Contents

List of Illustrations | ix

Foreword by Heather Falk | xi

Preface | xiii

Acknowledgments | xvii

Abbreviations | xix

Introduction | 1

1   The Doctrine of Original Sin: From Antiquity to the Present Day | 14

2   The Inheritance of Sin: Augustine's Transgenerational Doctrine | 31

3   The Transmission of the Soul as a Plausible Mechanism of Transgenerational Sin | 48

4   Epigenetics: A Key Scientific Window into Augustinian Theology | 62

5   Addiction: A Modern Reflection of Original Sin | 92

6   Original Sin and Addictive Propensity: Augustine's Theological Perspective | 109

Epilogue: Quantum Epigenetics and the Path of
    Sanctification | 135

Bibliography | 139

List of Illustrations

Figure 1. The Cycle of Addiction | 96
Figure 2: The Cycle of Redemption | 127

Foreword

WHERE FAITH MEETS SCIENCE, there are few explorations as profound and thought-provoking as the one contained within this book. *Genes of Eden* stands as a bridge across these two domains that many have perceived as insurmountably distant. Yet, Luman has approached this intersection with a rare blend of humility, intellectual rigor, and reverence for both tradition and discovery.

At the heart of *Genes of Eden* lies a question that has stirred minds and hearts for centuries: What does it mean to inherit the consequences of human choices? For early Christian theologians, like Augustine, this question birthed the doctrine of original sin, a concept that captured the transmission of a fallen nature across generations. Today, in a different vocabulary, we speak of transgenerational inheritance, epigenetic markers, and the molecular impact of life's burdens on future generations.

Luman has done the heavy lifting to help us understand addiction, a plight that has haunted humanity across millenia and cultures. Addiction, as he shows, is not merely a behavioral flaw or moral failing but can be understood as a deep-seated biological propensity, inherited in ways both overt and hidden. By uniting Augustine's understanding of humanity's bound will with epigenetic research on addiction, *Genes of Eden* invites us to reconsider the very nature of sin, struggle, and redemption in light of modern science.

The brilliance of this work is not merely in making theological concepts accessible to scientists or biological insights relatable to theologians; it is in drawing these together into a cohesive narrative of hope. Luman's chapters on TET1, JMJD3, and other molecular pathways provide a granular look at how trauma and addiction may indeed pass through generations. Yet, he does not leave readers in the realm of despair. Instead, he traces the possibility of healing, change, and renewal—a Cycle of Redemption that reflects the ancient promise of grace.

To the reader, *Genes of Eden* offers more than just information. It offers a perspective through which to understand addiction and human suffering with compassion and understanding. It is a text for those who grapple with personal struggles or walk alongside loved ones battling addiction; it is for theologians, scientists, and anyone who seeks to understand how we are shaped by the past while holding onto the potential for transformation.

Luman's work is a rare gift to those of us who dare to ask complex questions, who wrestle with the tension between what is seen and what is believed. In *Genes of Eden*, you are invited into that wrestling, to consider that our struggles may be both rooted in ancient faults and capable of being healed in the present. This book does not give all the answers, but it offers a foundation on which to seek them—a synthesis of science, faith, and the unyielding hope that renewal is possible.

As you turn these pages, I invite you to enter the garden of inquiry Luman has tended here. You may leave, as I did, with a profound sense of awe and a deeper understanding of what it means to be human.

Heather Falk, *Lost in My Mind* series

Preface

AT THIS JUNCTURE IN MY LIFE, as both a scientist and a pastor, I have come to a profound realization: the convergence of epigenetics and theology holds the key to unraveling the complexities of addiction. In *Genes of Eden*, I invite you to explore a new paradigm—a connection between early theology's treatment of original sin and the emerging field of epigenetics. My aim is to explore the mysteries of our genetic design, uncovering how the world around us influences not just our thoughts and bodies, but the very essence of our soul and spirit. With over three decades in the biopharmaceutical industry and years as a pastor, chaplain, and professor of cell and molecular biology, I wrote this book to address the struggles of those battling addiction by exploring how the theological implications of epigenetics can offer new understanding and hope. At the heart of *Genes of Eden* is a commitment to grounding its insights in Scripture, supported by theological and philosophical perspectives spanning from Origen to Kierkegaard. At the core of this venture is the question: Can the logic of science and the wisdom of theology come together to shed light on the mysterious control of addiction? Can addictive behavior be understood through the empirical rationalism of science, combined with divine revelation and metaphysical truths accessible only through faith? These questions form the foundation of *Genes of Eden*, grounding its exploration of the science-theology dialogue.

My journey into the science-theology arena that led to *Genes of Eden*, began with a fundamental question: How does the theological concept of original sin manifest within the human genome? Inspired by a background in both the science of molecular biology and theological ethics, I sought to bridge these disciplines, by investigating the relationship between genetic predispositions and behaviors, such as addiction, with the current science of epigenetics. By understanding how environmental factors influence gene expression, which is essentially what the science of epigenetics is, I examined how these external factors might affect addictive behavior. Inspired by renowned theologian Amos Yong, whose work bridges science and theology, I welcome his call for scientists and theologians to engage one another with a spirit of self-criticism, openness, and dialogue. His approach, grounded in these principles, provides the following encouragement:

> To engage theological authenticity in the encounter with science is to present our ideas self-critically, provisionally, and dialogically. Given the current state of the sciences, what can we say theologically? Of course, theology itself will be critical of any scientific claims which are metaphysical, or which exceed the limitations that science imposes upon itself. Yet good science sheds light on the nature of the world that God has created. While the overarching framework will continue to shift and expand, given the advances of science, there is reason to think that genuine truth is attained along the way, even if such truth will need to be further refined in light of future discoveries. The "science of holiness" therefore ought not to be avoided. Assuming we retain a robust yet appropriately humble theological foundation, science should not scare us off.[1]

Building on Yong's insightful comments—and with no intention of "scaring off" theologians—I have taken up the challenge to explore where science and theology meet. Advances in epigenetics offer a unique perspective for addressing theological questions about human nature and behavior. The study of epigenetics has revealed new possibilities for understanding changes in gene function,

---

1. Yong, "Sanctification, Science, and the Spirit," 52.

including those linked to cancer, autism, aging, and behavior. Epigenetic changes that include DNA methylation, histone modifications, and non-coding RNAs, influence how genes are expressed without altering the DNA sequence itself. These mechanisms can silence or activate genes to form complex traits and behaviors. Environmental factors like diet, stress, toxins, and psychological influences play a key role in defining how our genome impacts behavior—especially when it comes to addiction. Current research shows that addiction's epigenetic roots stem from the relationship of genetic, environmental, psychological, and, as this book explores, theological factors—all of which profoundly shape behavior. For those battling addiction, this understanding shifts the focus from personal blame to an inherited vulnerability—one born from the complex interaction of nature and nurture, both scientific and theological.

In recent years, studies have highlighted the role of trauma as a root of addictive behavior. Childhood trauma, for instance, can cause lasting changes in brain circuitry, increasing addiction risk. Children exposed to neglect or inconsistent caregiving are more likely to develop dysfunctional coping strategies, including substance abuse, to alleviate emotional pain. Theologically, this raises deep questions about sin and suffering: If behavior is defined by inherited predispositions, how do we understand original sin? According to Augustine, the main theologian referenced in this book, original sin represents humanity's inherited brokenness—a tendency towards selfishness and disobedience due to Adam's fall. This doctrine claims that each generation inherits both spiritual and moral inclinations towards sin. However, what if this sinful nature has not only spiritual, but also epigenetic foundations? Can epigenetic changes in behavioral genes, which lead to predispositions to harmful behaviors such as addiction, be inherited in a way that parallels the theological idea of original sin?

My overall aim in *Genes of Eden* is to address these questions and present a new paradigm for understanding addiction. This approach combines a unique perspective on epigenetics with the theological doctrine of original sin and the more recent theology of affordances, first described as, "what the environment offers the animal, what it provides or furnishes, either for good or ill," by psychologist James Gibson (1979).2 From Gibson's definition, affordances are the possibilities for action that the environment provides to living beings. The invitation to engage with creation, relationships, and Scripture can be understood through the concept of affordances, which are real, everyday expressions of divine grace that create opportunities for human growth and well-being. In the struggle against addiction, affordances provide a path from bondage to redemption.

Although many sciences are viewed as reductionistic, epigenetics provides insights that bridge both scientific and theological perspectives. Epigenetics also offers new pathways for intervention and resilience, by showing how changes in the expression of behavioral genes can occur. As Augustine observed, as will be discussed throughout *Genes of Eden*, human nature embodies both the burden of sin and the hope of grace. With this foundational truth, and the power of transformative faith complemented with science's evidence-based insights, *Genes of Eden* presents a spiritual and biological pathway to confront addiction, one of society's greatest challenges.

---

2. Gibson, "Ecological Approach to Visual Perception," ch. 8.

Acknowledgments

HAVING COMPLETED *GENES OF EDEN*, a journey of discovery, challenge, and growth, I am profoundly aware that it would not have been possible without the support, guidance, and encouragement of many individuals.

First, I would like to express my deepest gratitude to my wife, Pamela, and my family, all of whom have been invaluable sources of encouragement and support.

I am deeply grateful to my exceptional graduate supervisor in theological ethics at the University of Aberdeen, Brian Brock, whose guidance provided the foundation for my thesis, which became the basis for this book. Additionally, I extend my heartfelt thanks to my supporting reviewer, John Swinton, whose insights brought a profound pastoral perspective to the subject of addiction. Also, to Nuala Booth, a valued friend and steadfast source of support, who served as my PhD supervisor in biochemistry at the University of Aberdeen three decades before this work.

For my colleagues, Milton Karahadian, Steve Faivre, Jeff Salamat, Farhad Amiri, Haig Bozigian, Peter Voorhees, David Shirley, Brian Brodersen, Mihretu Guta, Amos Yong, Dennis and MaryAnn G Wilson, Gordon and Rene Debeever, Jeff and Julie Rodriquez, Willy Baptista, Bill Trok and David Sathiarag, thank you for engaging with my ideas, asking thoughtful questions, and inspiring me to think deeply. Your insights have enriched this work in countless ways. I am especially indebted to those who have graciously shared their expertise and wisdom. Your contributions have shaped the interdisciplinary dialogue at the heart of this book.

To my developmental editor, Charlie Collier, and the team at Wipf and Stock Publishers: your guidance, expertise, and meticulous attention to detail have been invaluable. Thank you for believing in this project and bringing it to fruition.

To those who have walked alongside me in faith, study, and life at University of Aberdeen and Calvary Chapel Costa Mesa, thank you for your prayers and encouragement. You have reminded me of the hope and grace at the core of this work.

Finally, I want to honor the countless individuals who have faced the challenges of addiction or stood by loved ones in their struggles. Your stories of courage and resilience have been a major source of inspiration in shaping *Genes of Eden*.

All this to say, this book is the result of a collective effort, and I am profoundly humbled and grateful to each of you who have contributed to its creation, and above all, my Lord Jesus Christ.

Abbreviations

CITED WORKS OF AUGUSTINE

C.Jul.	Answer to Julian *Contra Julian*
C.Jul.imp.	Unfinished Work in Answer to Julian *Contra Julian Opus Imperfectum*
Civ.Dei.	City of God *De Civitate Dei*
Conf.	Confessions *Confessiones*
Trin.	On the Trinity *De Trinitate*
Gn.Litt.	The Literal Meaning of Genesis *De Genesi ad Litteram*
nupt.et conc	On Marriage and Concupiscence *de Nuptiis* et *Concupiscentia*
Pecc.Mer.	On the Merits and Forgiveness of Sins and Infant Baptism *De peccatorum meritis et remissione et de baptismo parvulorum ad Marcellinum*

OTHER ABBREVIATIONS

AA	Alcoholics Anonymous
A, C, G, T	nucleotides adenine (A), cytosine (C), guanine (G), and thymine (T)
ARTS	Action, Reaction, Thinking Smart
BDNF	brain-derived neurotropic factor (protein)
BDNF	brain-derived neurotropic factor gene (genes are always italicized)
CH_3	methyl functional group
CRF	corticotropin-releasing factor
CRHR1	corticotropin-releasing hormone receptor 1 gene
CpG	Cytosine-phosphate-Guanine dinucleotide pairs
DB	disruptive behavior
DBD	DNA binding domain
DNA	deoxyribonucleic acid
DNMT1	DNA methyltransferase enzyme T1
DNMT3A	DNA methyltransferase enzyme 3A
DNMT3B	DNA methyltransferase enzyme 3B
DRD2	dopamine D2 receptor D2 gene
F_1	first filial generation
F_3	third filial generation
GABRA2	alpha-2 subunit of the GABA-A receptor gene
HPA	hypothalamic-pituitary-adrenal axis
IGF-1	insulin-like growth factor 1

ABBREVIATIONS

IGF2	insulin growth factor gene
JMJD3	histone demethylase; also known as KDM6B (lysine demethylase 6B)
KOR or *OPRK1*	kappa opioid receptor gene, often referred to as *OPRK1*
KRAB	Krüppel-associated box proteins, a family of transcriptional repressors
lncRNA	long non-coding ribonucleic acid
miRNA	micro ribonucleic acid
MLANA	melanoma antigen gene
mRNA	messenger ribonucleic acid
NA	Narcotics Anonymous
NR3C1	glucocorticoid receptor gene
NYP	neuropeptide Y gene
RNA	ribonucleic acid
RNA pol	ribonucleic acid polymerase
siRNA	small (or short) interfering ribonucleic acid
SLC6A3 (DAT1)	solute carrier family 6 member 3, encodes dopamine transporter gene
SLC6A4 (5-HTTLPR)	solute carrier family 6 member 4, encodes the serotonin transporter gene
TET	ten-eleven translocation (TET) enzymes
TET1	ten-eleven translocation 1 (part of the TET family of dioxygenases)
ZFP57	zinc-finger protein 57

Introduction

BOTH THEOLOGIANS AND SCIENTISTS have faced challenges in reconciling modern scientific phenomena with Christian doctrine. One of the greatest challenges has centered around the theories of the origin of life, particularly the conflicting perspectives of scientific abiogenesis and theological intelligent design. At the center of these controversies, recent breakthroughs in molecular biology have revealed a wealth of knowledge about the molecule at the heart of all living systems—deoxyribonucleic acid, or DNA. Within the exquisite maze of atoms that compose DNA, biologists have found that this double helical molecule is composed of four unique molecular units referred to as nucleotides or base pairs. Depending on their orientation, a specific sequence of base pairs, referred to as *genes*, can be located on ~23,000 regions of each DNA molecule.

Genes are the essential elements of DNA that code for the principal molecules, proteins, that compose a living being. DNA is found in every living cell on this planet. It is a template from which all living systems are constructed. Solving the structural sequence of the genes, specifically the human genome, has revolutionized the field of genetics. Moreover, understanding how genes function, or how they are expressed, has unveiled the extraordinary field of epigenetics, which describes how environment influences the function of genes. It is the scientific understanding of epigenetics that has provided a unique window for several fields of study, such as neuroscience and cancer research. Although the study of epigenetics has been centered in the biological sciences, other disciplines have profited from an understanding of how genes operate.

One particular field is theology, which is at the core of this book. Theologians may now be capable of proposing mechanisms for some perplexing theological issues, such as transgenerational sin. Theological doctrines such as sanctification, eschatology, pneumatology, and hamartiology may now be able to dialogue with science by critically appropriating insights from epigenetics. Within the fascinating field of epigenetics, scientists and theologians together now have a unique opportunity to deepen their exploration of complex issues like addiction and trauma. By bringing together the principles of epigenetics and theological ethics in treating addiction, we can uncover new, complementary perspectives that offer more sustainable paths to recovery.

Of the numerous stories of the addiction cycle, John C's experience typifies the life of unrelenting craving—essentially a mental imprisonment. However, as his story unfolds below, he found that the power of God, within himself, enacted a desire to change, to explore the path of recovery and develop resilience:

> For decades, John C., a follower of Christ, wrestled with the relentless grip of addiction. What began as an attempt to escape the pain of a broken childhood spiraled into a powerful force that seemed impossible to overcome. He tried everything—rehab programs, therapy sessions, and sheer determination. Each time, he believed this time would be different. But no matter how hard he tried or how long he stayed sober, the addiction always returned, often with even greater intensity. One night, after another crushing relapse, John sat alone, surrounded by the chaos his addiction had created. The weight of failure and despair was unbearable. In that moment of desperation, he cried out to God, something he hadn't done in years: "Lord, I can't do this anymore. I need Your help." The next day, while searching for a distraction, John found an old, dusty Bible he hadn't opened

in years. As he flipped through its pages, his eyes fell on 1 Cor 10:13, "No temptation has overtaken you except what is common to mankind. And God is faithful; he will not let you be tempted beyond what you can bear. But when you are tempted, he will also provide a way out so that you can endure it." The verse hit him like a lightning bolt. Could it be true? Could God really help him find a way out of the endless cycle of cravings and despair? John wasn't sure, but he knew he had to try. He began to read the Bible daily, clinging to 1 Cor 10:13 as if it were a lifeline. When the cravings resurfaced, he recited the verse aloud, using it as a shield against the urges. He also took a bold step and joined a small group at his church—a community of believers who embraced him with grace and understanding. They didn't judge him for his struggles but instead walked alongside him, offering prayer, encouragement, and accountability. It wasn't a quick or easy journey. There were days when the cravings felt insurmountable, but slowly, John noticed a change. The addiction, which had once seemed like an unbeatable enemy, began to lose its grip. He realized he wasn't fighting alone anymore. Through prayer, Scripture, and the support of his group, John discovered a strength beyond his own. Temptation still came, but it no longer had the same power. Each time it reared its head, John turned to God's promises and the people who had become his spiritual family. The battle wasn't over, but for the first time, he wasn't carrying the weight by himself. In his weakness, he had found God's strength, and it was enough.

John's journey is far from finished, but he now lives in the hope of God's faithfulness, knowing that the way out is always within reach when he leans on God. Today, John C. is addiction-free, living a life he once thought was beyond his reach. He knows the urges may return, but he also knows how to overcome them. Armed with faith, Scripture, and a community of believers, John C. has found the strength that had eluded him for so long. His story is not one of perfection, but of transformation—of a man who, after decades of struggle, finally found freedom through the power of God's grace.

Theological approaches, as demonstrated by John C., view addiction as a multi-dimensional issue, encompassing physical, psychological, social, and spiritual aspects. By addressing all these dimensions, individuals like John C. may find a more comprehensive path to recovery. By providing individuals with a sense of identity, hope, and a reason to change their addictive behavior, the process of transformation that renews the heart and mind can provide a greater source of strength and motivation. One of the major success factors for John C. was his participation in a Christian community. As John C. discovered, community involvement is a major aspect of theological approaches in that a community can provide a strong support network, accountability, and particularly a sense of belonging. This form of encouragement from like-minded individuals can be beneficial during the recovery process.

Rooted in a moral and ethical framework, theological approaches draw on scriptural values and principles to guide decision-making and help individuals avoid relapse. These principles complement a comprehensive assessment of an individual's medical, psychiatric, social, and addiction-related factors. Yet, beyond these social and psychological considerations lies a profound question: What spiritual and biological processes contributed to the transformative changes John C. experienced as he broke free from the grip of addiction? At the intersection of faith and biology, we must ask: What molecular events occurred in John's DNA to facilitate this transformation? Were specific epigenetic changes involved, and if so, which genes were affected? Can these genes be identified, and their expression monitored, using both theological insights and scientific methodologies? Moreover, do these two approaches—faith and science—align, offering a harmonious understanding of such profound change?

Over the past two decades there has been a heavy reliance on the potential of brain science to explain the cause, development,

INTRODUCTION

and clinical course of addictive behavior.[1] Brain science is often considered reductive as it focuses primarily on understanding human behavior and mental processes through biological mechanisms at the expense of other contributing factors. Human behavior and cognition are complex phenomena influenced by a multitude of interacting factors, including social, psychological, cultural, and environmental aspects. A brain-science model that solely focuses on neurological mechanisms may overlook the subtle details among these factors, thereby oversimplifying the understanding of human experiences.[2] Reductive models often seek to identify direct cause-and-effect relationships between biological processes and behavior, disregarding the complex, bidirectional nature of these interactions. Human behavior arises from a combination of genetic, biological, psychological, and environmental factors, and attributing it solely to neurobiology overlooks the multi-dimensional nature of behavior.[3] As objective as brain science is, the subjective experiences and meanings individuals attribute to their thoughts, emotions, and behaviors are often neglected, particularly with spiritual experiences.[4] This subjective aspect is an essential part of human existence that cannot be fully captured by biological explanations alone.

Neuroscience, while rapidly advancing, still has significant gaps in its understanding of brain function and its relationship to behavior. Relying solely on a brain-science model can lead to an incomplete understanding of human behavior and may overlook the contributions of other relevant fields, such as psychology, sociology, and in particular, theology. Reductive models that emphasize biological determinism may undermine the recognition of human agency and the capacity for self-determination. By overlooking the influence of personal choices, intentions, and societal contexts, a reductive approach may limit the significance of individual

---

1. Leshner, "Addiction Is a Brain Disease," 45–47.
2. Cacioppo, "Bisected Brain," 5–6.
3. Panksepp, "What Is Basic about Basic Emotions?" 387–96.
4. Haidt, "Righteous Mind," 285–318.

autonomy and personal responsibility.[5] By contrast, my approach in this book integrates multiple levels of analysis—including biological, psychological, social, and cultural factors—and theological perspectives, which represents a more comprehensive understanding of human nature.

Pharmaceutical treatments alone for the treatment of addiction, as I have observed, have failed to integrate the potential for families, communities, and other resources of enrichment to develop resilience in individuals who are suffering from drug addictions.[6] However, within the last decade drug-addiction researchers have become interested in studying the emerging field of epigenetics—a discipline that investigates an individual's environment and experience, specifically one's volitional experience.[7] Apparently, those who are addicted to a given substance due to repeated drug consumption have altered their DNA landscape within their brain in a regional and cell type–specific manner. By regulating DNA-related processes, drug-induced epigenetic alterations contribute to aberrant cellular function that drives drug-addiction pathogenesis. Understanding this mechanism by targeting key drug-induced epigenetic modifications within the brain may lead to novel interventions that could alleviate the persistence, and often relapse, of drug-seeking behaviors.

It has been demonstrated that prolonged stress, or even a single traumatic experience, influences the epigenetic landscape of the brain in ways that can lead to addictive behavior. With that in mind, the question arises as to what really is *epigenetics?* The most recited definition of epigenetics is *the modifications in gene expression that occur without changes to the underlying DNA sequence.* These modifications, however, can be very diverse and are influenced by various environmental factors, such as stress, trauma, diet and even people! The epigenetic changes that result from prolonged stress or a traumatic experience that may lead to addictive behavior are alterations of the genes expressed in brain

5. Lilienfeld, "Psychological Treatments that Cause Harm," 53–70.
6. Geschwind, "Meeting Risk with Resilience," 129–38.
7. Cloud, "DNA Isn't Your Destiny," para. 6.

INTRODUCTION

cells, specifically, without altering the underlying DNA sequence in the cells.[8] Why is this so profound? Interestingly, with no changes in the DNA sequence, or no mutations in the DNA sequence, it has been demonstrated that the epigenetic changes that occur are also reversible. The reversibility of epigenetic changes refers to the fact that alterations in epigenetic *marks*, such as DNA methylation, can be modified or reversed under certain conditions. Unlike changes to the underlying DNA sequence, which are generally considered permanent mutations, epigenetic modifications can be responsive to environmental cues and experiences. The various environmental factors include lifestyle, diet, stress, and exposure to substances. Changes in these environmental factors can lead to modifications in DNA methylation (or epigenetic marks). For example, adopting a healthier lifestyle, reducing stress, or undergoing certain therapies can impact epigenetic patterns and potentially reverse or modify existing epigenetic changes associated with certain conditions or behaviors.[9]

Epigenetic modifications are not fixed but exhibit a degree of variation or *plasticity*. Environmental exposures or even choices people make can mark or remodel the structure of DNA at the cellular level, and consequently, this may affect the whole organism. Epigenetic plasticity allows for potential alterations in gene expression patterns, particularly in a gene called the brain-derived neurotropic factor (*BDNF*) gene, which plays a major role in neuroplasticity. Certain medical interventions, such as epigenetic therapies, have been shown to modulate epigenetic marks and potentially reverse or modify epigenetic changes associated with specific conditions. For instance, behavioral interventions, such as mindfulness-based practices, have been associated with changes in DNA methylation and histone modifications.[10] Additionally, pharmacological agents targeting epigenetic processes, like DNA methyltransferase inhibitors or histone deacetylase inhibitors,

---

8. Cloud, "DNA Isn't Your Destiny," para. 19.
9. Waterland and Jirtle, "Transposable Elements," 5293–300.
10. Robertson, "DNA Methylation and Human Disease," 597–610.

have shown promise in modifying epigenetic marks and potentially influencing gene expression.

Further, spiritual experiences, such as prayer, fellowship with other Christians, and meditation on Scripture may also be associated with changes in epigenetic marks. Reversibility of epigenetic changes can even extend beyond an individual's lifetime, resulting in transgenerational epigenetic marks. Some epigenetic modifications can be altered during development and across generations.[11] Certain environmental exposures or interventions during critical periods of development across generations can modify epigenetic patterns, potentially leading to lasting changes in gene expression that can be reversible or even heritable.[12] The extent and duration of the reversible changes may also depend on individual variation and other factors.

Understanding the epigenetic mechanisms involved in addictive behavior, and whether they can complement or supplant other treatment strategies, is currently at the forefront of addiction research. Due to the changeability of epigenetic mechanisms that mediate addiction, several social and psychological approaches to recovery have been undertaken. However, one of the key disciplines that has not been included in the investigations of addictive behavior and epigenetics is theological ethics. Human freedom and biological correlates of epigenetic changeability with reference to original sin may provide a sound rationale for the causality of addictive behavior as a disorder or symptom of original sin.

Paul's letter to the Romans served as a foundational text for theologians who later articulated the doctrine of original sin, one of whom was Aurelius Augustine (354–430 AD), Bishop of Hippo, who was one of the four Fathers of the Latin Church (Augustine, Jerome, Ambrose, and Pope Gregory I). Each of these individuals played essential roles in developing Christian theology and shaping church doctrine. As a central, even revolutionary, figure within Christianity and Western thought, Augustine's views

---

11. Roth, "Lasting Epigenetic Influence of Early-Life Adversity," 760–69.

12. Skov-Jeppesen, "Changing Smoking Behavior and Epigenetics," 1565–75.

INTRODUCTION

on the doctrine of original sin were so profound that they were considered by some to be one of the root causes of theological division between the East and the West. Augustine produced volumes based on the apostle Paul's definitive statements on original sin, bringing theological as well as physiological perspectives into the discussion. In his writings, he explained sin as the *bondage of the human will*, which is highly suggestive for reflecting on the phenomenon of addictive behavior, in that sin and addiction are both voluntary and yet beyond the immediate control of a free will. Having this bondage to sin, or what Augustine refers to as *carnal concupiscence* and *constitutional fault*, are strikingly similar to the characteristics of addictive behavior. Understanding the connections between these two dispositions warrant an investigation into the causation and potentiation of sin and how these characteristics strongly resemble not only the susceptibility to addiction, but also the persistent mental state of craving.

Although Augustine lived at a time when biological science was at its infancy stage, particularly the field of genetics, he never makes it clear exactly how we were "in Adam," nor could he. Yet he did make attempts to comment on how sin could be inherited and even be transgenerational, by postulating that,

> Some sort of invisible and intangible power is located in the secrets of nature where the natural laws of propagation are concealed . . . in the loins of the father.[13]

A logical inference from these phrases is a reference to what we now know of as the science of genetics, which is the study of inheritance. As we understand genetics as the study of DNA and the structure and function of genes, the focus of this book is not on the *structural* composition or sequence of the genes, but on the how the genes *function*, or how they are expressed, which is explained by the relatively new science of epigenetics. It is within the context of the science of epigenetics that the implications of inherited sin can be explored.

---

13. Augustine, *C.Jul.imp.* IV.104.

Augustine argued that the original sin of Adam has been passed down intact to every member of humanity and cannot be alleviated by any act of the will because inherent sin in the human being comes with birth.[14] This is one of the hallmarks of Augustine's account of original sin. Although challenging, Augustine considers sin as a disease for which we are all responsible. As Alan Jacobs explains:

> Many of us would also agree that sin, like the more communicable diseases, transfers easily to other people; few of us have strong immunity to its ravages. But we would also agree that the affliction of disease is not moral in character. Although it is possible to act in such a way that one becomes more prone to illness, surely there is no sin in being ill. Disease, we tend to agree, happens to us; sin is what we do. Yet it is just this simple and familiar distinction that Augustine—drawing on the passages in Genesis and Romans—denies. In his account, an infant who has not "acted" and is "not at fault" has nevertheless, somehow, broken a covenant with God.[15]

In Augustine's view, original sin introduces a fundamental flaw in human nature, resulting in a state of spiritual and moral bondage. This perspective aligns with the concept of addiction as a manifestation of disordered desires and the loss of self-control. Original sin resulting in brokenness and disordered desires in human nature parallels the distorted reward-pathways observed in addiction. Augustine's emphasis on the need for grace and divine intervention in overcoming the effects of original sin can be related to the role of epigenetic mechanisms in addiction recovery. Epigenetic modifications are reversible and can be influenced by environmental factors, including interventions and therapeutic approaches. Just as Augustine asserts the importance of divine grace in redemption, the concept of epigenetic plasticity suggests that interventions, such as Christian disciplines of prayer, fellowship,

---

14. Augustine, *Civ.Dei.* XIV.7.
15. Jacobs, *Original Sin*, 133–40.

INTRODUCTION

and study of Scripture in addition to environmental changes, can modify epigenetic marks and potentially restore gene expression patterns associated with godly behaviors.

While Augustine's view of original sin does not directly account for the specific mechanisms of epigenetic changes, it provides a theological basis from which to understand the consequences of addictive behavior and the potential influence of epigenetics on addiction-related genes. By considering the inherent weakness and spiritual struggle in Augustine's view, we can draw connections to the biological processes observed in addiction and the potential role of epigenetic changes in contributing to addictive behaviors. The emerging field of epigenetics is currently addressing many of the complicated challenges in several fields of interest, including biology, psychology and sociology. However, epigenetics has yet to be introduced into the theological arena, and for the reasons described above epigenetics is one of the key sciences that should be given an audience within theology.

Beginning with an overview of Augustine's doctrine of original sin, this book will focus on the heritability and transmissibility of original sin in terms of two aspects: common guilt and constitutional fault. This section will then be followed by a discussion of the biology of epigenetics and how environment shapes our genes. Next, I will examine how epigenetic changes can be heritable and transmissible and even changeable—that we are not necessarily the victims of our genes. These ideas will then be coupled with heritability and transmissibility of original sin, how these can be evidenced by the expression of certain genes. Bringing together an understanding of the changeable nature of epigenetic mechanisms and the theology of Augustine's original sin can shed light on the root cause of addictive behavior. This section will lead to a discussion on how the redemptive act of Christ, empowered by the Holy Spirit, leads to recovery.

With this understanding, recovery can be sustained through a life that is committed to responsibility, which is the essence of resilience. Even with ongoing struggles with addiction, an individual committed to growing in their walk with Christ can develop the freedom of the will and the desire to honor God rather than self, as exemplified by John C. By integrating a biological context with the power of the Holy Spirit, those ensnared by addiction can be equipped with resources to experience the freedom that God has designed for them. Current physical, psychological, and philosophical treatments of addiction continue to face the challenges of relapse. A successful recovery strategy can promote a resilient life, marked by a sense of belonging—especially within Christian programs that emphasize forgiveness and gratitude.[16] Having an attitude of forgiveness, both towards oneself and others, can alleviate guilt and promote healing. It is, therefore, a key objective of this book to understand addiction within the science-theology context of epigenetic inheritance of original sin. Understanding how behaviors and environment can cause changes in the way genes function, the basis of epigenetics, opens a unique window from which to consider the potential for altering the sinful condition of humans. In the context of sin, epigenetics can be seen as a modern parallel to the theological doctrine of sanctification—the process by which individuals are gradually made holy or being transformed into God's image. In other words, just as epigenetic changes occur as the result of environment, susceptibility to addictions could be diminished through a combination of therapeutic interventions, environmental changes, and spiritual growth. It is important to note, however, that these ideas must be grounded in Scripture, as Augustine attests:

> In matters that are obscure and far beyond our vision, even in such as we may find treated in Holy Scripture, different interpretations are sometimes possible with prejudice to the faith we have received. In such a case, we should not rush headlong and firmly to take our stand on one side that, if further progress in the search for truth justly undermines this position, we too fall with it. That would be to battle not for the teaching of Holy Scripture but for our own, wishing its teaching to conform to ours,

---

16. Krentzman, "Gratitude and Addiction Recovery," 301–9.

INTRODUCTION

whereas we ought to wish ours to conform to that of sacred scripture.[17]

If we view sinfulness as an inherited *trait*, epigenetics suggests that it is not necessarily a permanent or unchangeable state. Epigenetics provides a compelling possibility of altering the sinful condition. It suggests that while sin may be deeply ingrained, it is not beyond the reach of change, particularly when an individual commits to a life devoted to God and receiving his grace, mercy, and forgiveness. This perspective offers hope to those struggling with addiction by suggesting that they are not bound by their genetic makeup alone. Instead, they have the capacity for transformation, not only through medical and psychological support but also through spiritual and personal development. This approach is in line with the idea that individuals can overcome deeply ingrained behaviors and tendencies, whether they are of a spiritual or physiological nature, such as addiction or even the transgenerational effects of trauma.[18]

It is imperative, however, to understand the limitations as well as the possibilities associated with epigenetics, that this book is not based on reductionist biological determinism. By understanding and leveraging epigenetic principles presumably associated with theological doctrines such as original sin, individuals may find a pathway to freedom from addiction. The concept of transformation through epigenetics could also extend to addressing susceptibility to addictions, a condition often linked to both a genetic predisposition and environmental factors. Addiction, viewed as a paradigm of sinful behavior in Augustine's context, is deeply rooted in both biology and experience—yet, mercifully, it is not necessarily immutable.

---

17. Augustine, *Gn.Litt*, 1.18, para. 37.
18. Yehuda, "Public Reception of Putative Epigenetic Mechanisms," 1–7.

1

The Doctrine of Original Sin

From Antiquity to the Present Day

THE DOCTRINE OF ORIGINAL SIN, a central tenet of Christian theology, has undergone substantial development over the centuries. Its origins trace back to the apostle Paul, who laid the groundwork for later theological interpretations by emphasizing the disobedience of Adam and its ramifications for humanity (Rom 5:12–21). Early Christian theologians began to wrestle with these ideas, setting the stage for later interpretations. John Chrysostom, for example, was an influential Archbishop of Constantinople, and he played a major role in shaping early Christian reflections on sin and human nature. Renowned for his expressive homilies and commentaries, particularly on Genesis and Romans, Chrysostom explored the consequences of Adam and Eve's disobedience, yet he did not develop the doctrine of original sin nearly to the extent that his contemporary, Augustine, would later do. Instead, he focused on the moral and spiritual corruption stemming from the Fall, emphasizing individual responsibility and free will over inherited guilt.[1] Augustine of Hippo, the central character of this book, profoundly shaped the Western understanding of original sin in his exhaustive works that included *Confessions*, *City of God*, and *On the Merits*

---

1. Chrysostom, *Homilies on Genesis*, 1–17.

and Forgiveness of Sins, and the Baptism of Infants. These texts and many others became central to Catholic and later Protestant theology and laid the foundation for the doctrine by stating that all humans inherit Adam's sin and guilt, leading to a corrupted nature from Adam to all humanity.

Following Augustine, the doctrine of original sin evolved significantly, re-formulated by later theologians and historical contexts, as summarized in the comprehensive works of Pelikan[2] and McGrath.[3]

Pope Gregory the Great was one of the earliest to expand on Augustine's ideas of original sin. In his *Moralia in Job*, he highlighted humanity's inherent fallenness and the need for divine grace, clearly revealing Augustine's influence. This theological course continued into the medieval era, where figures like Anselm of Canterbury emphasized that original sin created a debt to God, a debt that only Christ's atoning sacrifice could satisfy. Thomas Aquinas later expanded the discussion, focusing on original sin as the loss of original justice and its transmission to all humanity, having a major influence on Catholic theology.

The Protestant Reformation brought a new wave of interpretation, particularly Martin Luther's emphasis on the total depravity of human nature and John Calvin's integration of original sin into his doctrines of predestination that underscored humanity's absolute dependence on God's grace. In response, the Catholic Church reaffirmed traditional teachings during the Council of Trent, maintaining the necessity of baptism for the remission of original sin. This confirmation contrasted with the evolving Protestant views, causing a major theological divide.

Post-Reformation debates were diverse, with individuals like Francisco Suárez and Cornelius Jansen revisiting Thomistic and Scholastic ideas. Suárez focused on grace and human nature, while Jansen's interpretation, heavily influenced by Augustine, emphasized predestination and human corruption, leading to the Jansenist movement. Simultaneously, in the Lutheran tradition,

---

2. Pelikan, "Christian Tradition," ch. 2.
3. McGrath, *Christian Theology*, chs. 1–3.

Johann Gerhard emphasized the effects of original sin on human nature, reinforcing the need for divine grace. In America, Jonathan Edwards advanced the Reformed understanding, stressing inherited corruption and the necessity of transformative grace.

The Enlightenment brought a shift, with a more optimistic view of human nature leading some to reinterpret original sin. Friedrich Schleiermacher viewed it as a psychological and existential condition—an alienation from God rather than a transmitted guilt. This reinterpretation, focusing on human consciousness, paved the way for nineteenth- and twentieth-century liberal theology. John Henry Newman engaged with the doctrine in relation to sacraments and justification, while Karl Barth's theology reframed original sin as a universal condition requiring redemption through Christ. Modern theologians like Paul Tillich and Hans Urs von Balthasar took similar existential approaches, emphasizing a discordance with God and the necessity of divine grace for restoration. Reinhold Niebuhr saw original sin as humanity's tendency toward pride and self-interest, a concept that reflected the social challenges of his time.

The Catholic Church, through the Second Vatican Council, became more ecumenical by integrating modern thought with traditional doctrine. Although the Council did not explicitly address original sin, its themes of human dignity and holiness provided a more optimistic view of human nature. The 1992 Catechism of the Catholic Church continued to uphold the traditional teaching, highlighting the consequences of original sin and the importance of baptism. In the late twentieth century, John Paul II's *Theology of the Body* and Joseph Ratzinger's writings emphasized original sin's implications for human dignity, sexuality, and redemption, linking it closely with the transformative power of grace.

Today, the doctrine remains a subject of discussion and reinterpretation. Contemporary theologians such as N. T. Wright critique individualistic interpretations, favoring a narrative centered on Israel and Christ's role in redeeming Adam's fall. James Cone and Miroslav Volf bring new perspectives, viewing original sin in terms of social justice, systemic racism, and reconciliation, emphasizing grace's role in overcoming human divisions.

For both Protestants and Catholics, debates persist as believers seek to reconcile traditional teachings with modern scientific and philosophical insights, emphasizing the doctrine's lasting complexity and significance. While many denominations continue to uphold traditional interpretations, there is a growing interest in reinterpreting the doctrine in light of contemporary insights into human nature, psychology, ethics, and even scientific advancements. With this in mind, greater opportunities to explore the natural world and its Creator opens the door to richer dialogue between scientific inquiry and theological perspectives.

In the following sections, selected essays from the recent volume *Augustine and Science*[4] are discussed as a preface to the examination of Augustine's doctrine of original sin, followed by an introduction to the science of epigenetics as it relates to addiction. These essays present dialogues between scientists and theologians, where they examine the strengths and limitations of their respective fields. While complete agreement may remain tenuous, the dialogue enhances each perspective, aiming for a shared understanding of God's creation while acknowledging the inherent limits of human knowledge.

AUGUSTINIAN SCIENCE

The doctrine of original sin, deeply entrenched in Christian theology, offers profound insights into the nature of creation and providence. Rooted in the Genesis narrative and expounded upon by Paul the apostle, the doctrine of original sin was developed extensively by Augustine. At the core of this doctrine lies the belief that all humanity inherits a sinful nature because of Adam's disobedience.[5] This foundational narrative, found in the first three chapters of Genesis, portrays God as the Creator who brought the

---

4. Doody, *Augustine and Science*, 1.
5. Augustine, *Gn.Litt.* I.V.

universe into existence *ex nihilo* (out of nothing) and established the natural laws that govern it. Drawing from narratives in the books of Genesis and Romans, Augustine expressed a clear distinction between the period of creation and the era of divine intervention or *providence*. During the creation period, God produced the universe according to his divine plan and established the natural laws that govern its operations. Also during this era, various life forms emerged, each reflecting the Creator's wisdom and goodness. In contrast, the era of providence encompasses the present age, where all life forms interact within the confines of the natural laws established during creation. While God's creative activity is not confined to the past, his providential intervention sustains the ongoing order of the world. This distinction demonstrates the continuity of God's involvement in creation while recognizing the completed nature of the original act of creation.

In the recent volume *Augustine and Science*,[6] several essays explore the twenty-first-century perspective on *Augustinian science*, with particular attention to avoiding anachronisms, which are ideas or events placed in a time period where they do not belong (often because they emerged from a later era). Recognizing that Augustinian science would have been known as *natural philosophy*, it is evident from his works that Augustine was centered primarily in politics and literature and was on the fringe of natural philosophy. Contemporary scholars, however, have considered his thoughts in later scientific debates regarding the age of the earth and the theory of evolution. As these topics continue to be hotly debated, works from Augustine, such as *De Genesi ad Litteram* (Literal Meaning of Genesis), have been recognized as essential references for various sides of debates.

Augustine has been, and continues to be, a pivotal figure whose theological and philosophical ideas have transcended centuries, influencing not only religious thought but also the foundations of Western philosophy. Augustine's reflections on creation, time, and knowledge are the major themes of the essays included in the referenced text, *Augustine and Science*. Among the

---

6. Doody, *Augustine and Science*.

prominent thought leaders in science and theology, particularly in patristic theology, is Ernan McMullin, who while not a direct contributor to *Augustine and Science*, is an influential voice whose work on critical realism is often engaged in the volume's discussions. Ernan McMullin developed a *critical realism*—a philosophical stance that supports Augustine's view of creation, recognizing both the reality of the external world and the fallibility of human perception in understanding it.[7] McMullin explores Augustine's profound contemplation on the nature of creation as an act of divine will. For Augustine, as stated previously, creation is *ex nihilo*, an expression of God's omnipotence and freedom. This idea is crucial for McMullin, who sees the intelligibility and order of the universe as reflective of a Creator. Augustine's belief that the created world is good, ordered, and knowable sets the stage for McMullin's argument that science, in exploring this order, is fundamentally a theistic endeavor.

McMullin draws directly from Augustine's concept of "the two books"—the book of nature and the book of Scripture—both of which are seen as revelations from God. Augustine taught that these two sources of truth, while different in form, are ultimately harmonious because they come from the same divine source. McMullin accepts this Augustinian perspective and argues that scientific and theological knowledge should not be seen as opposing forces, but as complementary ways of understanding reality.

In this view, any apparent conflict between science and theology is a result of human misinterpretation, not a fundamental incompatibility. McMullin's critical realism is essentially Augustinian in that it acknowledges the limits of human knowledge. Augustine was well aware of the finite and fallen nature of human understanding, as he emphasized the need for humility in the pursuit of truth. McMullin applied this perspective to his philosophy of science, arguing that while science can uncover real truths about the world, these truths are always tentative and subject to revision. This agrees with Augustine's idea that our knowledge of God's creation is always incomplete, and that faith and reason must

---

7. Allen, *Ernan McMullin and Critical Realism*, ch. 4.

work together to approach a fuller understanding. McMullin then applies Augustine's ideas to contemporary debates in the science-theology dialogue, particularly in areas like cosmology, evolution, and the interpretation of Genesis.[8] The narrative explains how McMullin uses Augustine's thoughts on time and creation to engage with modern cosmology, especially the Big Bang theory.

Augustine's assertion that time itself was created by God helps McMullin argue that scientific explanations of the universe's beginning are not in opposition to the doctrine of creation. For Augustine, there was no "before" creation because time began with the universe. McMullin applies this to counter the idea that the universe's temporal beginning implies a conflict with eternal divine creation. Instead, he suggests that scientific descriptions of the universe's origins are a way of uncovering the mechanics of God's creative act. It is from this perspective that McMullin engages with evolutionary theory and how it is informed by Augustine's flexible interpretation of Genesis. Augustine, acknowledging the limits of human understanding, recognized the potential for the biblical text to be allegorical or symbolic and was open to non-literal readings that could accommodate different interpretations of creation. McMullin extends this Augustinian approach to even include evolutionary theory, suggesting a reading of Genesis that is not necessarily at odds with the account of evolution. By following Augustine's lead, McMullin argues that the "days" of creation could be metaphorical, representing long periods rather than literal 24-hour days. This interpretation could allow a reconciliation, or continued dialogue, between the doctrine of creation and the theory of evolution, positioning God as the ultimate cause behind the natural processes. McMullin's exploration of divine intervention is an area where Augustine's influence is particularly strong. Augustine's belief that the universe is contingent—that is, it exists because God willed it, not out of necessity—provides McMullin with a theological basis for understanding the natural world's inherent unpredictability and order. McMullin argues that the laws of nature and the processes of evolution could even be

---

8. Allen, "Augustine and the Systematic Theology of Origins," ch. 1.

expressions of God's providential care, where God's purpose could be realized through natural means. This view allows McMullin to see scientific explanations not as threats to divine action but as descriptions of how God's will is manifested in the world. Understanding this perspective requires an intellectual humility in the science-theology dialogue.

McMullin is fully aware of Augustine's recognition of the limits scientific knowledge, that scientists and theologians are unable to fully grasp the infinite nature of God or the entirety of his creation. McMullin promotes a position of openness and humility in both scientific and theological pursuits, acknowledging that both disciplines are continually evolving and that our understanding is always partial. This Augustinian humility, according to McMullin, is essential for a productive dialogue between science and theology, where both sides are open to learning from each other and to revising their views considering new evidence and insights.

Augustine's theological insights with McMullin's critical realism establish a foundation from which to engage science and theology. Augustine's ideas about creation, time, and the nature of knowledge provide the philosophical and theological basis for McMullin's approach, allowing him to understand current scientific developments while remaining committed to his Christian faith. It is with critical realism, aided by Augustine, that McMullin can engage with contemporary scientific theories in a way that respects both the integrity of science and the truths of Christian theology.

Augustine emerges from this encounter not just as a historical figure but as a vital source of wisdom for modern discussions on the origins of the universe, life, and humanity. McMullin's application of Augustinian thought to the science-theology dialogue demonstrates how ancient insights can still provide valuable guidance in contemporary intellectual challenges, offering a path forward for those who seek to reconcile faith with the discoveries of modern science. With this desire in mind, McMullin has been credited with the word *consonance*, as a way of pursuing the dialogue of reconciling faith and modern science.

One of the key essays of the *Augustine and Science* text is Paul Allen's paper, "Augustine and the Systematic Theology of Origins."[9] Allen explores how Augustine's theological ideas can inform contemporary discussions on the origins of the universe, life, and humanity. Allen takes readers on a journey through Augustine's thought, showing how the early church father wrestled with questions of creation, time, and the nature of humanity—topics that are still relevant in modern debates between science and religion. The story begins with Augustine, a deeply philosophical theologian living in the late Roman Empire, who grappled with questions that continue to challenge modern minds. Allen reviews how Augustine's understanding of creation goes beyond a literal reading of the Genesis text, as was noted above. He believed that God created the world *ex nihilo* and that creation was an unfolding process rather than a single, instantaneous event. As Augustine considered the nature of time, he proposed that time itself was part of God's creation. For Augustine, there was no "before" creation, because time began with the universe. This idea, Allen suggests, provides a valuable perspective for modern discussions on the origins of the cosmos, particularly in the context of Big Bang cosmology.[10]

Allen then shifts the focus to Augustine's thoughts on human origins. Augustine believed that humanity was created in the image of God, but he also acknowledged the symbolic and allegorical elements in the Genesis account. Augustine was open to the idea that the "days" of creation could represent longer periods, aligning with the notion of a gradual development of life. Throughout the narrative, Allen emphasizes that Augustine's theology of origins is not in conflict with science but can be seen as complementary. Augustine's willingness to engage with the best knowledge of his time, while maintaining a firm theological foundation, offers a model for how modern Christians might approach questions of origins today.[11] Allen concludes with a reflection on the implications of Augustine's thought for contemporary systematic theology.

9. Allen, "Augustine and the Systematic Theology of Origins," 9–26.
10. Allen, "Augustine and the Systematic Theology of Origins," 14.
11. Allen, "Augustine and the Systematic Theology of Origins," 18–20.

He argues that Augustine's approach encourages a dialogue between theology and science, where both can enrich each other. Allen sees Augustine's legacy as a call for humility and openness in the face of complex questions about the universe and our place in it. In the end, Augustine is seen not just as a historical figure, but as a living conversation partner for those who seek to reconcile faith and reason in the quest to understand the origins of everything.

Allen's paper touches on the influence of Ernan McMullin's concept of "Augustinianism" within the context of theological discussions on origins.[12] McMullin, a philosopher of science, was deeply engaged with Augustine's thought and sought to apply it to contemporary dialogues between science and religion. Allen makes key comments on McMullin's adaptation of Augustinianism, advocating for a dialogue where theological insights and scientific discoveries inform one another rather than exist in isolated spheres. Allen points out that McMullin's Augustinianism encourages a similar approach, one that is open to the findings of science while remaining rooted in theological tradition.[13] McMullin was inspired by Augustine's intellectual humility, and he warned against the dangers of trying to make science fit theological doctrines too rigidly or vice versa. Allen underscores this point, noting that McMullin's Augustinianism is marked by a careful and patient approach, recognizing that both science and theology are ongoing, evolving conversations, exemplified by the current dialogue generated in this book, *Genes of Eden*.[14]

In another essay from *Augustine and Science*, "Science: Augustinian or Duhemian?", Alvin Plantinga explores two contrasting views on the relationship between science and religion— "Augustinian" and "Duhemian" perspectives, named after St. Augustine and Pierre Duhem, respectively.[15] or Plantinga, Augustinian science represents a model in which theological convictions are allowed to inform scientific inquiry, rather than requiring a

12. Allen, "Augustine and the Systematic Theology of Origins," 21.
13. Allen, "Augustine and the Systematic Theology of Origins," 22.
14. Allen, "Augustine and the Systematic Theology of Origins," 24.
15. Plantinga, "Science: Augustinian or Duhemian?" 153.

stance that is independent of theological commitments. Instead, Augustine suggests that religious beliefs, particularly Christian ones, can and should influence scientific inquiry. Proponents of this view argue that science is a human endeavor deeply connected to our broader worldview, including religious commitments. In *The Literal Meaning of Genesis*, Augustine cautions against carelessness in hypothesizing about creation accounts, whether framed by creation science or evolutionary theory, as stated earlier":

> That would be to battle not for the teaching of Holy Scripture but for our own, wishing its teaching to conform to ours, whereas we ought to wish ours to conform to that of Sacred Scripture.[16]

From the Augustinian perspective, the interpretation of Genesis must balance humility with theological conviction. In contrast, Duhemian Science, named after the physicist and historian Pierre Duhem, advocates for a religiously neutral approach to science. Duhem argued that science should be conducted without reference to religious or metaphysical beliefs, focusing solely on empirical data and logical reasoning. This approach maintains that science and religion operate in separate domains and should not influence each other. Plantinga critiques the Duhemian view, suggesting that it is difficult to maintain complete neutrality in science, as scientists often work within a model influenced by their broader worldview.[17] He leans toward the Augustinian approach, arguing that it is more realistic and that religious beliefs can provide valuable insights and motivations in scientific exploration. However, Plantinga also emphasizes the importance of intellectual humility and openness to different perspectives in the pursuit of scientific knowledge. To reinforce this perspective, Plantiga refers to Ernan McMullin's concept of "consonance," which is an approach to understanding the relationship between science and religion.[18] Unlike models that view science and religion as either in conflict

---

16. Augustine, *Gn.Litt.* I.18.37.
17. Plantinga, "Science: Augustinian or Duhemian?" 160.
18. Plantinga, "Science: Augustinian or Duhemian?" 165.

or entirely separate, McMullin proposes that they can be in a state of *consonance*, where they are in agreement without necessarily overlapping or directly influencing each other. McMullin's idea of consonance complements Alvin Plantinga's discussion of the Augustinian and Duhemian perspectives by offering a middle ground. While Plantinga critiques the Duhemian ideal of religious neutrality in science and leans toward the Augustinian model that integrates religious belief into scientific inquiry, McMullin's consonance suggests a different kind of relationship: one where science and religion maintain their distinct domains but are not in opposition. In McMullin's view, consonance occurs when scientific findings and religious beliefs are compatible or when they can be interpreted in ways that do not conflict. This does not mean that one directly informs the other but that they can collaborate in a way that is intellectually rewarding.

Plantinga appears to value McMullin's consonance as compatible with his Augustinian model. However, Plantinga's Augustinian perspective would push further, advocating for an even more integrated approach where religious beliefs actively shape scientific inquiry. One could argue that McMullin's consonance provides a practical perspective for those who are more aligned with the Augustinian model but operate within a predominantly secular scientific community. It allows for a personal integration of faith and reason without demanding that the scientific community adopt a religious foundation. This approach might be seen as a more feasible way to balance the aspirations of Augustinian science with the realities of a pluralistic world.

The following chapters of this book explore the concept of theology and epigenetics, seeking to position it within McMullin's idea of consonance. This will provide a valuable perspective on the complex relationship between science and religion, and within Plantinga's critique of strict Duhemian neutrality, both science and theology are recognized as having distinct but interrelated roles. The well-known saying, "In essentials, unity; in non-essentials, liberty; in all things, charity," attributed to Augustine, reflects a timeless wisdom expressed across most Christian traditions, from

Augustine to John Wesley and beyond. This aphorism is helpful when considering the complex relationship between science and theology. However, this call to unity raises two vital questions: "What defines the essentials?" and "Who holds the authority to determine them?" These questions are particularly significant when considering Augustinian science, as discussed by Plantinga, who engages with Duhem's ideal of religious neutrality in science. The Augustinian model that combines faith with scientific inquiry, based on McMullin's consonance discussed in the next section, invites the relationship where science and theology maintain their distinct domains but are not in opposition.

AUGUSTINIAN CONSONANCE

Drawing upon the foundation built from our engagement with *Augustine and Science*, this book explores the concept of theological epigenetics, which also attempts to reconcile scientific theories with theological principles. Two other essential essays from *Augustine and Science* by Patrick Richmond and James Spiegel reinforce the objectives of theological epigenetics, highlighting the compatibility between scientific understanding and Augustinian theology.

Patrick Richmond, in his essay "An Augustinian Perspective on Creation and Evolution," presents an Augustinian framework for harmonizing contemporary scientific views with theological beliefs, demonstrating a consonance between the two.[19] Richmond emphasizes Augustine's nuanced approach to the book of Genesis, cautioning against the pitfalls of interpreting it in a strictly literal sense. Instead, Augustine understood Genesis as both a historical account and a theological narrative, as stated previously, by interpreting the 'six days of creation' metaphorically rather than as literal 24-hour periods. This interpretation could allow for a non-literal reading of the Genesis creation narrative, which, as Richmond points out, opens the door to compatibility with modern scientific

19. Richmond, "Augustinian Perspective on Creation and Evolution," 181–94.

theories like evolution.[20] Richmond highlights Augustine's belief in a God who is both transcendent and immanent, continuously sustaining and governing creation. Augustine spoke of creation as both a singular divine act and as an unfolding process under providence. This perspective agrees with the idea that God could have created the universe with the inherent potential for development, allowing natural processes, such as evolution, to occur within the divine plan.

Augustine's understanding of divine causality does not conflict with evolutionary change but can encompass it as part of God's creative activity. A critical aspect of Augustine's theology is his concept of time, which Richmond discusses in depth. In his *Confessions*, Augustine argues that time itself is a part of creation, beginning with creation and distinct from God's eternal nature. For Augustine, all of history—including the evolutionary history of the cosmos—is present to God in a single, timeless vision. This view challenges the rigid separation between creation and evolutionary development, suggesting instead that all temporal events, including evolution, unfold within God's eternal plan.

Richmond argues that Augustine's theological approach offers valuable insights for contemporary Christians grappling with scientific accounts of evolution. Augustine's belief that God created the world with inherent causal powers supports the idea that natural processes, like evolution, could be instruments of divine creation. Additionally, Augustine's flexible interpretation of Genesis provides a framework for engaging with scientific theories without abandoning core theological commitments. While Augustine's thought is adaptable to modern understandings of evolution, Richmond notes that Augustine did not anticipate the specific details of Darwinian evolution. Augustine's focus was more on the metaphysical and theological aspects of creation than on empirical science. Therefore, to avoid scientific anachronisms, Augustine's views should not be used to directly support contemporary evolutionary biology. However, Augustine's emphasis on

---

20. Richmond, "Augustinian Perspective on Creation and Evolution," 183.

the compatibility of faith and reason provides a sound justification for dialogue between science and theology.

Patrick Richmond concludes that Augustine's theology—through its interpretation of Scripture, understanding of God as Creator, and flexible conception of time—offers a constructive approach for reconciling Christian faith with evolutionary science. Augustine's perspective encourages believers to view the natural world, including its evolutionary development, as an expression of God's creative will. An Augustinian perspective, Richmond suggests, can create a consonance between science and religion, where evolutionary theory is not seen as a threat to Christian belief but as an opportunity for continued dialogue. As sciences such as epigenetics are unveiling greater knowledge of genetics, for example, we can continue to strive for a deeper understanding of God's creation.

Complementing Richmond's perspective, James Spiegel's essay "Augustine, Evolution, and Scientific Methodology" delves into Augustine's theological and philosophical ideas, particularly his understanding of faith and reason and their relevance to the scientific method.[21] Spiegel explores how Augustine's approach provides a valuable method for engaging with scientific inquiry while upholding theological commitments. Spiegel explains that Augustine saw faith and reason as complementary ways of knowing, rather than contradictory. According to Augustine, all truth ultimately comes from God, whether discovered through natural reason or divine revelation. Augustine believed that reason could aid in understanding faith, and faith could guide reason, yet there are limits of human knowledge. This balanced approach, according to Spiegel, supports scientific inquiry without undermining theological beliefs. Augustine taught that human cognition is limited and fallible and that true knowledge comes from reason and revelation. While Augustine acknowledged the validity of empirical observation, he also stressed that human senses and reason are prone to error.

---

21. Spiegel, "Augustine, Evolution, and Scientific Methodology," 195–210.

This *epistemological humility* aligns with the scientific method's principles of skepticism and continuous revision, where knowledge is always provisional and subject to refinement. According to Spiegel, Augustine is flexible in his approach to interpreting Scripture, especially when it seems to conflict with empirical evidence. That interpretations of Scripture should adapt to accommodate well-established scientific observations. When a literal reading of Scripture conflicts with scientific evidence, Augustine believed it was likely due to a misinterpretation of the Scripture. This principle supports a congruent relationship between scientific discovery and theological interpretation, particularly as Spiegel argues that Augustine's thought offers a supportive foundation for the scientific method.

Augustine recognized the importance of empirical observation and rational analysis in understanding the natural world, viewing it as a reflection of God's order and wisdom. This belief in the intelligibility of creation suggests that the natural world can be studied systematically and rationally, a core principle of the scientific method. Augustine's view that all truth is God's truth means that the pursuit of scientific knowledge is not inherently in conflict with faith. However, Augustine cautioned against the overreach of science, warning against knowledge that "puffs up" or leads to pride. Spiegel rightly instructs that Augustine believed the ultimate purpose of knowledge is to deepen our understanding and love of God. This perspective is particularly relevant today, reminding us that scientific knowledge should be pursued with humility and recognition of its limitations.

In agreement with Spiegel, I believe that it is of great value to support Augustine's approach to faith, reason, and scientific inquiry. It offers valuable guidance for contemporary discussions on evolution and scientific methodology by promoting an integrative perspective that avoids both scientific reductionism and theological dogmatism. By acknowledging the limits of human reason, maintaining openness to empirical discovery, and interpreting Scripture with flexibility and humility, Augustine demonstrates consonance as he provides a model for us all—essentially a *constructive consonance*, particularly with the challenges posed by evolutionary theory and modern science.

Building on the dialogue between science and theology, the next chapter explores Augustine's doctrine of original sin, providing a detailed examination of how this foundational Christian belief intersects with contemporary scientific insights, particularly in the field of epigenetics. Augustine argued that sin is not merely an individual act, but a condition inherited from the first human beings, affecting all of humanity. Modern epigenetics provides a fascinating biological perspective for understanding this concept, suggesting that certain traits and behaviors can be inherited across generations through mechanisms beyond Mendelian genetic transmission. This emerging science opens the possibility that the transgenerational inheritance of sin could find an analogy in the biological process by which environmental influences affect gene expression across multiple generations.

Augustine's view of sin as a pervasive, insidious force closely resembles the understanding of addiction in contemporary science. Addiction can be seen as an inherited vulnerability that is both biological and spiritual in nature. Examined through the perspectives of original sin and epigenetic inheritance, the subsequent chapters will reveal how addiction itself is not merely a personal failing but a manifestation of the deeper, more profound brokenness that Augustine describes as a condition passed down and perpetuated through generations. This blending of theology and science presents an extraordinary opportunity to explore the mechanisms of epigenetic inheritance and how they can deepen our understanding of human limitation, sin, and ultimately, redemption.

2

The Inheritance of Sin

Augustine's Transgenerational Doctrine

THE CHRISTIAN DOCTRINE OF original sin refers to the dramatic consequences of the first sin that was committed by Adam and Eve when they succumbed to the temptation that was craftily presented by the serpent as described in the book of Genesis chapter 3. According to the traditional understanding of original sin, the whole human race and future generations were infected with a general condition of sinfulness in the Fall of Adam and Eve from grace. Original sin *is not personal* in that it is not the consequence of personal choice or personal failure to act, yet it *is personal* in the sense that everyone is personally subject to the effects of original sin. Atonement for original sin requires the redemptive act of Jesus Christ. By believing that his death and resurrection are sufficient for the forgiveness of one's sin, evidenced by baptism that washes away original sin, one is then saved from the consequences of sin.

Augustine of Hippo was a prominent philosopher, theologian, and bishop who lived from 354 to 430 CE. Born in modern-day Algeria, he is considered one of the most influential figures in Western Christianity. Augustine was initially influenced by the Manichaean religion and later by Neoplatonism. However, his life took a significant turn when he converted to Christianity in 386

AD after a spiritual crisis. He dedicated himself to the Christian faith and became a prolific writer and defender of Christian doctrine. His most famous work is *Confessions*, an autobiographical account of his early life and spiritual journey. Additionally, his treatise *The City of God* is a significant contribution to Christian philosophy and political thought. Augustine's teachings have had a profound impact on Christian theology, shaping doctrines such as original sin, divine grace, and the concept of the just war. His work has left a legacy in both religious and philosophical circles, making him one of the most influential thinkers in the history of Christianity.

Augustine ultimately turned to Christianity, as described in his *Confessions*, after hearing God's voice direct him to Rom 13:14: "But put on the Lord Jesus Christ, and make no provision for the flesh, to fulfill its lusts." Among Augustine's most significant teachings are his detailed writings on original sin that were circulated in the fifth century. Although Augustine's writings were lauded by the Catholic Church, Pelagius, a British monk, challenged his views because of the moral laxity prevalent in the church. Pelagius believed that Augustine's theology did not stress man's participation in and responsibility for his sanctification and would encourage further moral laxity. Pelagius thought that men were considered innocent in God's eyes until they reached an age of accountability. He held the view that human nature was essentially good, and that sin was a personal act that could not be inherited in any way. Furthermore, Adam's sin established a pattern for sin, and he set us all a bad example. For Pelagius, original sin was therefore *socially transmitted*. In Pelagius's view, sin is an act of human freedom, and humans remain free to make moral choice—sin only comes about through moral choice; it cannot be passed down from one's ancestors.[1]

In addition to Pelagius, the Eastern Church and even some contemporary Western theologians have never embraced Augustine's doctrine of original sin. Their major criticism is that original sin does not see man as a free moral agent who has maintained his

---

1. Bray, *Augustine on the Christian Life*, 39.

free will after the fall of Adam. Although they agree that sin has wounded humanity, they do not believe that Augustine's doctrine of original sin fully explains how humans participate in Adam's sin or how this inherited condition influences the exercise of free agency.[2] While Scripture does not explicitly address human participation in Adam's sin, the evidence that original sin has spread to all of humanity is clearly stated in Rom 5:12: "Therefore, just as through one man sin entered the world, and death through sin, and thus death spread to all men, because all sinned." If this Scripture is also referring to the guilt of Adam's sin, as some critics have also countered, then the following Scriptures will also need to be accounted for, such as Rom 5:18: "So then as through one transgression there resulted condemnation to all men, even so through one act of righteousness there resulted justification of life to all men." And 1 Cor 15:22: "For as in Adam all die, so also in Christ all shall be made alive." In agreement with the critics, these passages may be understood as affirming Adam's sin as the origin of humanity's fallen condition. These Scriptures clearly show that Christ's death erased Adam's original sin. That salvation and justification bring an end to the guilty verdict associated with original sin. With justification is also positional sanctification, which is a one-time event that occurs at the moment of salvation, where believers are declared holy and righteous through their faith in Christ.

As positionally right with God, a believer lives a life of progressive sanctification, which is a continuous journey of spiritual growth and transformation as a believer is conformed more and more into the image of Christ throughout life.[3] Progressive sanctification is affirmed in 2 Cor 4:16: "For which cause we faint not; but though our outward man perish, yet the inward man is renewed day by day." As new creatures in Christ Jesus, even though we sin and are guilty of the sins that we commit in our *outward man*, we are being made holy in our *inner man*, progressively, day by day, by the work of the Holy Spirit.[4]

2. Bray, *Augustine on the Christian Life*, 45.
3. Heb 10:10; Heb 10:14; 1 John 1:9.
4. 2 Cor 5:7; John 16:8.

Another challenging criticism of original sin relates to infant baptism, as there is strong scriptural justification to show that children are innocent in God's eyes until they reach an age of accountability.[5] If the Augustinian view of original sin is to be believed, unbaptized or unsaved infants who die must suffer eternal damnation to maintain theological consistency with the doctrine, which is a serious and controversial consequence of this doctrine. Although the Pelagian doctrine of the social transmission of sin has been generally rejected, the way original sin is transmitted from one generation to the next remains an enigma for theologians. The idea of Augustine's *inherited sin* implies a Mendelian-type of "law of inheritance," which is similar to how eye color or other genetic traits are hereditary. If the social transmission of sin does not align with Scripture, then could sin be hereditary and sustained by the environment? Is it possible that heredity and environment work together, and that original sin is the sum of heredity and environment as explained by *epigenetics*? Might this be a way of understanding what Paul is referring to as the *old man* that is carried around with the believer—and why Paul encourages the believer to crucify the old man in Rom 6:6 ("Knowing this, that our old man is crucified with him, that the body of sin might be destroyed, that henceforth we should not serve sin")? How does original sin affect the believer's *old man*? Is this the battle that every believer struggles with, particularly those who are struggling with addictions?

The prolific writings on the doctrine of original sin by Augustine come from his personal issues with sexual desire as well as his refutations of the Pelagian theory that original sin occurred in Adam and Eve only, that it is not inherent in humanity. However, Augustine profoundly explains how original sin has been *propagated* in humans, that human DNA is tainted with a *materiality* of sin. This inference of the materiality of sin being propagated in all humans suggests that transgenerational sin could occur by an epigenetic phenomenon, which is explored in this book with relation to addictive behavior.

5. Deut 1:34–39; Is 7:15–16.

The doctrine of original sin, rooted in Genesis 3, was elaborated upon in Paul's letter to the Romans, then further expounded upon by Augustine, who taught that the sin of Adam and Eve has been passed down to every member of humanity. Accordingly, Augustine developed this concept of the origin of sin to state that sin cannot be remedied by any act of the will because sin is inherent in every human at birth. There is not only a spiritual separation from God, but also a separation evidenced physically and emotionally, that human beings since Adam inherit a propensity towards sin since humans are born in sin: "Behold, I was brought forth in iniquity, and in sin my mother conceived me" (Ps 51:5). Paul added that we are "by nature children of wrath" (Eph 2:3). And because of the fall of Adam and Eve, "all sinned in Adam" (Rom 5:12). Sin is, essentially, a turning away from God that causes disorder to the mind, body, and soul, so that every human being is in a psychological dilemma with one's self. This distorted desire, a *pathological wanting* for satisfaction, is a disorder Augustine associates with the sin of pride.[6] As such, Man "exchanged the truth of God for the lie and worshipped the creature rather than the Creator . . . " (Rom 1:25), as demonstrated by seeking fulfillment in possessions and ultimately power over all.

With a reputation for his supposed disdain of contemporary theologies of the body, Augustine described a condition of sexual reproduction before the fall, in which Adam's *member* was fully under the control of his rational will.[7] Only after the fall does lust take over, and the member gains a *mind of its own*. With his emphasis on disordered desire, Augustine presents an account of the image of God as found not in the flesh but in human rationality.[8] On a number of occasions, Augustine describes an enslavement to lust and sexual desire, particularly in his *Confessions*, with remarks that have led some to conclude that Augustine despised all things bodily.[9] However, Augustine saw uncontrolled, disordered desire

6. Augustine, *Civ.Dei* 14.7.
7. Augustine, *Civ.Dei* 14.7.
8. Augustine, *Trin.* 12.7.12.
9. Augustine, *Conf* 8.5.10.

as the real evil, which he called *carnal concupiscence*, and not the sexual organs themselves. Augustine describes sexual desire as his primary example of the disordering and disempowering effects of carnal concupiscence. He sees that our potential for good is challenged with our carnal concupiscence, citing Paul's declaration, "the good that I will to do, I do not do; but the evil I will not to do, that I do" (Rom 7:19). Augustine echoed the cries of Romans 7 by concluding that after the Fall, Man is *not able not to sin*, and that Man had *lost the liberty of the will*; through freedom man came to be in sin, but the corruption that followed as punishment turned freedom into *necessity*.[10] The sins of necessity are due to a corruption of nature, and every human is born with the *propensity* to sin.

Although there is no direct evidence that sin is transmitted in the genes, the Pelagian view, as countered by Augustine, wrongly claims that we have no inherited propensity to sin, but rather that everyone simply sins of their own free will. The Pelagian position was that Adam's sin affected Adam and only Adam and no changes occurred in humanity as such. That sin was limited to Adam only and was not transgenerational. Augustine argued, in contrast to the Pelagians, that man was created in a state of righteousness, in the image of God, yet he was not created *immutably* so, because it was possible for him to sin.[11]

Original sin, then, has man in a condition in which he is unable to refrain from sinning, with disordered desires that remain chained by his evil tendencies. Augustine reiterates that the freedom that remains in the will always leads to sin and in the flesh, that we are free only to sin, which is a freedom without liberty, and essentially a *moral* bondage. Augustine concluded that true liberty can only come from without, from the work of God on the soul, and we are completely dependent upon grace for our conversion. Accordingly, we are born into lives, relationships, and social structures distorted in all kinds of ways by sin. These are distortions to our beliefs, desires, affections, and wills so that as soon as we are capable of willing and acting ourselves, we do them in ways

10. Augustine, "Perfection of Human Righteousness," 4.9.
11. Gaul, "Augustine on the Virtues of the Pagans," 242.

distorted by sin. Our present condition of being *bound to sin* is inherited by each of us before we are ever in a position to choose it. Original sin does not "obliterate our capacity for willing and choosing, but rather *co-opts and diverts* it."[12] In this view, we are responsible or accountable to God for our present state of sinfulness even though we inherited it before we could ever choose it and are unable to avoid it.[13] In essence, Augustine's doctrine of original sin is that original sin is inherited, which means that all humans are essentially born in a process of *moving away from God*.[14] Sin is not simply evil; sin is a blameworthy evil for which one is responsible. Thus, original sin maintains that all human beings are born blameworthy of sin, born in a state of being *misrelated* to God.

A modern proponent of the doctrine of original sin, Reinhold Niebuhr, determined that original sin is empirically confirmed because it is everywhere around us and has been evidenced throughout world history.[15] Many pagans in antiquity were convinced that a great sin lay behind all the misery in the world.[16] Distorted human desires have been exemplified throughout the ages, as G. K. Chesterton describes:

> The ancient Master of Religion began with the fact of sin—a fact as practical as potatoes. Whether or not man could be washed in miraculous waters, there was no doubt at any rate that he wanted washing. But certain religious leaders . . . have begun in our day not to deny the highly disputable water, but to deny the indisputable dirt. Certain new theologians dispute original sin, which is the only part of Christian theology which can really be proved . . . Only with original sin can we at once pity the beggar and distrust the king.[17]

---

12. McFadyen, *Bound to Sin*, 16–18.
13. Messer, *Theological Neuroethics*, 96.
14. Greer, *Christian Life and Christian Hope*, 201.
15. Niebuhr, *Nature and Destiny of Man*.
16. Kung, *Great Christian Thinkers*, 85.
17. Chesterton, *Orthodoxy*, 16.

Augustine's explanation of original sin was centered on two major Scriptures by Paul to the Romans and the Corinthians. In Rom 5:12, Paul surmises that, "sin came into the world through one man, and death came through sin, and so death spread to all, *in whom* all have sinned." Augustine takes Adam to be the one *in whom* all have sinned, not to all *because* all have sinned. This idea of inheritance is reinforced in 1 Cor 15:22, "For as in Adam all died . . . ", which is understood by the contrast of Adam and Christ—the sinfulness in Adam is as righteousness that is solely found in Jesus Christ.

AUGUSTINE'S DOCTRINE: FIVE FOUNDATIONAL THEMES

Augustine's doctrine of original sin can be partitioned into five components, according to Couenhoven.[18] Beginning with the source of original sin, referred to as (1) *primal sin* in the garden of Eden, all human beings share in this sin because of our (2) *solidarity* with Adam, the progenitor of humanity. From birth, based on this solidarity with Adam, all human beings have an (3) *inherited sin*, which comes in two forms: (a) *common guilt* and (b) a *constitutional fault* of disordered desire and ignorance. As a result of the inheritance of these two forms, the human race suffers a (4) *penalty* because of sin—human powers are weakened, and humans will experience death. The consequences of the inheritance, both original sin and its penalty are (5) *transmitted* from generation to generation.

With numerous volumes written on original sin, what follows is a focused discussion of the most relevant to the inheritance and propagation of original sin. Of the primary texts cited for these two points, in addition to his *City of God* and debates with the Pelagians, is Augustine's final book, the *Unfinished Work in Answer to Julian*. This text provides insightful allusions to how original sin is inherited and transmitted from generation to generation, which

---

18. Couenhoven, "St. Augustine's Doctrine of Original Sin," 359–96.

are ideas that support the causal basis of addictive behavior based on epigenetic inheritance of origin sin.

Augustine's doctrine begins with the observation that all sin originates from pride, demonstrated by the desire to live by the rule of self.[19] Eating the fruit in the garden was an external expression of the sin that was already present.[20] Thus, the evil will that is already present, resulted in an evil act. The cause of sin did not arise in the flesh, but in the soul.[21] Adam and Eve were created to rest and trust in God, and this unimaginable act due to a distorted desire supports Augustine's concept of evil as deprivation. Sin is a fall from the good and is always irrational, as evidenced by an inherent deficiency and disorientation, so that only grace can challenge sinful desires.[22] Human nature was impaired after the fall and resulted in a discord between flesh and spirit (which became our new nature), and our intellectual and physical faculties were weakened.[23] How this impairment of human nature took place, and how our sin was contracted from Adam, is dependent on our human solidarity with Adam.

As head of the human race, Adam contained all future human beings within himself (with the exception of Jesus Christ, the second Adam, who was born of a virgin), and his choice to sin shaped the sin nature of those who proceeded from him. In Augustine's view, solidarity with Adam is more than a legal matter of a covenant between God and the human race. Adam, as our representative, had the power to ratify or break the treaty for all of us. Solidarity with Adam, for Augustine, is both social and ontological, as he describes:

> We were all in that one man, seeing that we all were that one man who fell into sin. We did not yet possess forms individually created and assigned to us for us to live in them as individuals; but there already existed the seminal

19. Augustine, *Civ.Dei* 12.6; *Gn.Litt.* 11.5.7.
20. Augustine, *Civ.Dei* 14.13, 42; *Gn.Litt.* 11.5.7.
21. Augustine, *Civ.Dei* 14.3, 13.
22. Augustine, *Civ.Dei* 12.6; 12.7, 9.
23. Augustine, *C.Jul.imp.* 5.20.

nature from which we were to be begotten . . . when this was vitiated through sin . . . man could not be born of man in any other condition.[24]

Augustine is suggesting that all human beings live double lives—he refers to our lives in Adam as a "common life" of souls not yet living separately. He also calls our lives in our own bodies' *individual* and *proper* lives.[25] Whether his theory is rooted in *traducianism* (related to the materiality of the soul) or to Platonic and Plotinian ideas of the preexistence of the soul, he affirms that our souls partake of the common life of the human race, in Adam, before they were enlivened in our own bodies.

From Augustine's perspective, the reason all human beings begin life as sinners is because human sinfulness is historically grounded, based on the primal sin of Adam and Eve and on human solidarity in Adam, i.e., all are now sinners because all were in Adam. The need to be redeemed by Christ is not because of our potential to sin, but because we have original sin. And all humanity is guilty before God: even those who never have a chance to will or act on their own have inherited sin because of our solidarity with Adam. Although Augustine never makes clear exactly how we were in Adam, his reflections on inheritance can be creatively interpreted in light of later genetic insights, even though he could not have anticipated Mendelian genetics. Remarkably, his musings contained insights that resonate with modern understandings of heredity, such as, "Some sort of invisible and intangible power is located in the secrets of nature where the natural laws of propagation are concealed", with the observation that original sin to have been, "in the loins of the father."[26]

From this passage, Augustine is suggesting that all humans existed in Adam's seed, which is an explanation that calls for elucidation. Even though this passage implies seminal existence in Adam, it is not clear as to whether he is inferring a *materiality*

---

24. Augustine, *Civ.Dei* 13.14.
25. Augustine, *C.Jul.imp.* 3.32.
26. Augustine, *C.Jul.imp.* 6.22.5.

THE INHERITANCE OF SIN

of sin, with reference to the traducian theory of the origin of the soul. Augustine never provides any details as to how we all existed in Adam, primarily because there were no concurrent theories of inheritance to make any reference to at that time. Had Augustine been aware of the biology of *epigenetic inheritance*, he may have considered a traducian materialistic origin of the soul. Due to his inability to even speculate on a *biological* solidarity in Adam, he speculates to his opponent Julian that *some sort of invisible and intangible power located in the secrets of nature where the natural laws of propagation are concealed* is at the root of original sin. From our current understanding of embryology, the natural laws of propagation occur within a zygote, a fertilized egg within the womb, which is a scientific concept that would not be known until almost 1,500 years later. To ratify his speculation, however, Augustine simply concludes his theory with the mandate, "if you cannot understand this, believe it."[27]

Augustine's idea for inherited sin may have originated from references in Tertullian and Origen.[28] Additionally, Ambrose, who was a contemporary of Augustine, referred to all of humanity as being *in Adam* with a type of hereditary sin being washed away at baptism.[29] However, Ambrose did not develop a theory of how guilt was inherited. Augustine did make a definitive observation that solidarity in Adam and inherited sin are both ontological and social by concluding that by one man sin entered into the world, and death by sin; and it passed upon all men, in which all have sinned.[30] This passage is an affirmation of Paul's conclusion that all have sinned because all men were in Adam when he sinned.[31] Augustine understood human persons as *dependent beings with porous selves*, who receive their identities from God and the other people around them.[32] Adam and Eve exemplified this at their

27. Augustine, *C.Jul.imp.* 4.104–5.
28. Tertullian, "Treatise on the Soul," para. 2.
29. Augustine, *Trin.* 1.3.
30. Augustine, *Pecc.Mer.* 3.14.
31. Rom 5:12.
32. Couenhoven, *Stricken by Sin, Cured by Christ*, 35.

creation, as God had imparted them with virtue, purity, and innocence.

From their inception, human agents are accountable for what they own, and that persons own their hearts and minds, which make them who they are as persons. Similarly, when Adam and Eve fell, they had ownership of their now fallen beliefs and desires. So even while they lacked control over their fall into sin, just as they lacked control over being created righteous, they did not lack accountability for being fallen. The goodness that they found within themselves, without having chosen it, paralleled Augustine's conception of the original sin into which all of Adam's race are born.[33] The original sin was participatory and human beings are imputed original sin because we were in Adam, and thus *sinned with him*. For Augustine, it is not our legal situation that makes us guilty, we are judged guilty because we are *already* guilty, by virtue of our *seminal participation* in Adam. We are judged and penalized for what Adam did because we participated, somehow, in the primal sin. Thus, Adam's sin and ours are one and the same.

To have participated in the primal sin, Augustine speaks of original sin as the guilt from our origin which was contracted by birth, which he refers to as *common guilt*.[34] This also refers to our solidarity with Adam, since we all were in Adam when Adam sinned, the guilt for his sin is ours from the moment of birth. The guilt of Adam's sin remains in us as a *stain*, until and unless it is forgiven in baptism, even when the one who personally committed the sin—Adam—has died.[35] The entire human race was lumped together with Adam when he sinned; as a result, the whole of mankind is a *condemned lump*, the stock being punished along with the roots.[36] The common guilt that results from the primal sin, however, is neither sin's punishment, nor a further sin; rather, it is the moral stain left on the human race by the primal sin itself. Augustine also speaks of original sin as an inherited state of

33. Couenhoven, *Stricken by Sin, Cured by Christ*, ch. 4.
34. Augustine, *C.Jul.imp.* 5.29.
35. Augustine, *C.Jul.imp.* 4.96, 116.
36. Augustine, *Civ.Dei* 21.12.

disordered desire and ignorance, a *constitutional fault* with which we are born.[37] According to this account, original sin is more than participatory, it is our own corrupted state—a state of disordered love and ignorance—which we have in our own proper lives, not the marginal existence we have in Adam's loins.

Augustine explains *inherited constitutional fault* as a turning away from the Creator. The corrupted human nature, rooted in distorted desire, is what Augustine calls *carnal concupiscence*. Although concupiscence is defined as desire in general, Augustine does not view all concupiscence as negative. Concupiscence in and of itself is not necessarily evil desire. Augustine sometimes uses the term *concupiscence* in a narrower sense, where he notes evil concupiscence as carnal, or fleshly, concupiscence. Thus, carnal concupiscence is a desire for things forbidden, and the actual desire for sin, a disobedience coming from ourselves and against ourselves.[38] Augustine's use of the word *carnal* may indicate the blaming of the body for sin. However, he states that the weakness of our post-fall existence dwells in both the flesh and the soul.[39] Augustine claims that when we say that the flesh hears, we mean that the soul hears by means of the ear.[40] Augustine explains carnal concupiscence as follows:

> The cause of carnal concupiscence is not in the soul alone, much less in the flesh alone. It comes from both sources: from the soul, because without it no pleasure is felt; from the flesh, because without it, carnal pleasure is not felt. Hence, when Paul speaks of the desires of the flesh against the spirit, he undoubtedly means the carnal pleasure which the spirit experiences from the flesh and with the flesh as opposed to the pleasures which the spirit alone experiences.[41]

37. Augustine, *C.Jul.imp.* 5.29.
38. Augustine, *nupt.et conc.* 2.9.22.
39. Augustine, *Gr.et pecc.or.* 1.12.13.
40. Augustine, *Gn.Litt.* 10.12.20.
41. Augustine, *Gn.Litt.* 10.12.20.

As stated, Augustine thinks it is an error to believe that sin is in the body only. Any sins one might attribute to the flesh reflect on the soul, because the latter has been given governance of the former.[42] Augustine recognizes, moreover, that sins are not particularly bodily problems: the devil has no body, but he still has many faults, especially pride and envy.[43] So carnal concupiscence is the disordering of the whole person, body and soul. Sexual desire, post-fall, exemplifies this disobedience with painful clarity for Augustine. Our sexual desires are not merely animal or biological; they reach our deepest inner being. Yet these desires come and go without the permission or direction of the conscious will, and they are not properly oriented towards higher goods. Instead, they resist and distort reason, according to Augustine:

> You say this, as if we say that concupiscence of the flesh surges up only into the pleasure of the sex organs. This concupiscence is, of course, recognized in whichever sense of the body the flesh has desires opposed to the spirit.[44]

Augustine also discusses how non-sexual desires, not only lust, stem from carnal concupiscence, including the desire to eat food simply for the taste (and not for nourishment) greed for money, and lust for power and domination.[45] Even though Augustine appears to focus on lust as the major cause of carnal concupiscence, he does write that sexual desire in the garden neither preceded the will nor went beyond it.[46] Even the sense appetite, Augustine called the senses of concupiscence to which he attributes positive or negative meanings.[47] The senses of the flesh are not in themselves desires, though they can arouse our desires with the sensory information that they supply to the rational soul.

42. Augustine, *Civ.Dei* 14.23.
43. Augustine, *Civ.Dei* 14.3.
44. Augustine, *C.Jul.imp.* 4.28.
45. Augustine, *C.Jul.* 4.14.65–74.
46. Augustine, *C.Jul.imp.* 2.122.
47. Augustine, *C.Jul.imp.* 4.27.

Augustine considered carnal concupiscence as *sin*, and not merely an evil, but a culpable and blameworthy evil. Disordered desire, as carnal concupiscence, is not only evil but produces guilt that must be forgiven, even if conscious consent is lacking.[48] Thus, Augustine's view that carnal concupiscence, even if it is not acted on, or out of, is itself sin. His repeated assertions that the guilt of concupiscence is forgiven in the rebirth that is baptism clearly imply that carnal concupiscence is itself sinful. Concupiscence of the flesh is not forgiven in baptism in such a way that it no longer exists, but in such a way that it is not counted as sin.[49] Prior to baptism, then, carnal concupiscence creates guilt, and this is sinful.

Post-baptism, Augustine affirms that carnal concupiscence remains in the baptized as a bad quality, like a disease. The baptized, too, have disordered desires. Their guilt, however, is forgiven, and Augustine suggests that concupiscence is thus no longer literally, but only figuratively, sin. He explains that this is what it means to be without sin, not to be guilty of sin.[50] For those who are baptized, concupiscence is not counted as sin, it remains evil, but its guilt is taken away.[51] It can, however, provide the occasion for personal sins, choices and actions that arise from our disordered desires, which is the reason why devout Christians continuously battle their evil desires. In response to this, Christians are commanded by Paul, "Let not sin reign in your mortal body to make you obey its lusts" (Rom 6:12).

Although confessing base desires, receiving forgiveness, and absolving guilt are personal responsibilities, Augustine believes that because of having solidarity with Adam, human beings have the potential to fundamentally harm other human beings, making them sinners before God. Human beings are thus vulnerable to the influences of others, even to the extent that their relation to others determines their standing before God. Yet God's sovereignty

---

48. Augustine, *C.Jul.* 6.16.49–50.
49. Augustine, *C.Jul.* 6.14.42.
50. Augustine, *nupt.et conc.* 1.26–9.
51. Augustine, *C.Jul.* 6.17.51–2.

prevails through the church, reconciling sinners in Christ's work of redemption.

From inherited sin that resulted from primal sin and solidarity in Adam, as a brief summary of the penalty of the primal sin, Augustine states that while the penalties of sin are a fitting punishment for sin, they are not sinful in themselves. They can provide reasons and occasions for sin, but since they do not necessarily lead to sin, they are evil without being intrinsically sinful.[52] The most prominent penalty of sin is mortality; the first couple was meant to live forever, but death is now the penalty of all who are born.[53] As this is not what God intended; but since the fall, all human beings are penalized by death. As in death, described by the ancient Israelites, as water being poured onto the ground, human weakness, loss of beauty, susceptibility to disease are ways in which we are in the process of dying even before we are dead.[54] For Augustine, such weakness plainly indicates that we live under a penal condition.

The final and fifth point of Augustine's doctrine of original sin is regarding the transmission of original sin, which is the most relevant to epigenetic inheritance. The transmission of original sin has the challenge, as Augustine recognized and was unable to resolve, of being connected with the various theories on the soul's origin and transmissibility. With this in mind, it became clear that the transmission of sin could be compared to the transmission of the soul. Three main theological theories on the origin and transmission of the soul are also relevant when considering how sin is transmitted: creationism, traducianism, and preexistence. The *creationism* theory states that the soul is specially created and is introduced into the body when the body is being formed, which is unlike *traducianism*, that the soul, like the body, is transmitted from one generation to the next in a natural manner. For

---

52. Augustine, *C.Jul.* 4.16.49.
53. Augustine, *Civ.Dei* 13.6; *C.Jul.imp.* 1.96.
54. Johnson, *Vitality of the Individual in the Thought of Ancient Israel*, 9.

preexistence, the soul predates the body and comes into the body at the appropriate moment.[55]

As suggested, it is from these theories of the soul that a mechanism for the transgenerational epigenetic inheritance of sin is proposed. The theological implications of various theories provide explanations for how environmental disturbances might affect both the soul and sin, and how these effects could be passed on to future generations. They also suggest mechanisms by which such influences might be inherited, proposing that both the soul and sin can be passed down from one generation to the next. Given this context, the next chapter explores the foundational works of early Christian thinkers such as Origen, Tertullian, Augustine, and Aquinas, as each of these theologians offer unique insights into the nature and transmission of the soul.

---

55. Pelikan, *Shape of Death*, 82–85.

3

The Transmission of the Soul as a Plausible Mechanism of Transgenerational Sin

ENGAGING THE WRITINGS OF early Christian theologians, one can begin to formulate a theory as to how original sin might be inherited alongside the soul. This exploration aims to integrate ancient theological perspectives with contemporary understandings of how environmental factors might influence the transmission of the soul, ultimately providing a greater understanding of the mechanism by which original sin is inherited.

ORIGEN: PRE-EXISTENCE OF THE SOUL

The concept of the preexistence of souls is a fascinating and complex topic that has been explored by various religious and philosophical traditions throughout history. Origen, an early Christian theologian from the third century, explored this idea as part of his theological foundation. Origen's proposition of the preexistence of souls emerged within the broader context of his theological system, which sought to reconcile Christian doctrine with philosophical thought, particularly Platonic philosophy. To understand Origen's perspective, based on his seminal text, *On First Principles*,[1] it is

---

1. Origen, *On First Principles*, 1, 1–8.

essential to grasp his interpretation of the nature of God, the cosmos, and the human soul. Origen's affirmation of God as eternal, transcendent, and immutable corresponds to a significant theological strand within Scripture. In his cosmology, Origen proposed a hierarchical structure of reality, with various levels of existence extending from God. This hierarchical framework encompasses both the spiritual and material realms, with each level reflecting differing degrees of perfection and proximity to God. To Origen, theological anthropology is unfolded in the concept of the soul's journey towards perfection and union with God. He conceived of the soul as inherently rational and immortal, possessing the capacity for moral agency and spiritual growth:

> In regard to the soul, whether it takes its rise from thy transference of the seed, in such a way that the principle or substance of the soul may be regarded as inherent in the seminal particles of the body itself; or whether it has some others beginning, and whether this beginning is begotten or unbegotten, or at any rate whether it is imparted to the body from without or not; all this is not very clearly defined in the teaching.[2]

However, Origen contends that the soul's journey is filled with challenges and obstacles, primarily the appeal of sin and the entanglements of the material world. In Origen's perspective, the preexistence of souls serves to account for the origin and destiny of the human soul. His proposition was that prior to the creation of the material universe, God fashioned the souls of all intelligent beings, endowing them with rationality and freedom. These souls existed in a state of intimacy with God, sharing in his divine life and purpose. However, some souls, driven by pride or curiosity, turned away from God and embarked on a downward course towards material existence. The descent of the soul into bodily existence represents a pivotal moment in Origen's cosmology. It marks the soul's separation from its divine origin and the beginning of its journey through the cycle of reincarnation or migration. Origen's

---

2. Origen, *On First Principles*, 1, 5.

concept of the soul's descent reflects Plato's concept of the fall from the realm of the Forms into the realm of *becoming*, where the soul becomes ensnared in the imperfections of the material world.

Origen's teaching on the preexistence of souls raises theological questions regarding the nature of divine intervention, human freedom, and the problem of evil. On one hand, Origen emphasizes the sovereignty of God as the creator and sustainer of all things, including the souls of intelligent beings. God's foreknowledge and benevolence encompass the entirety of the soul's journey, guiding it towards its ultimate reconciliation with divine goodness. On the other hand, Origen affirms the reality of human freedom and moral responsibility. The soul's decision to turn away from God and embrace material existence involves a radical exercise of autonomy, challenged with moral consequences.

Origen struggles with the tension between divine sovereignty and human agency, suggesting a connection between God's providential care and the soul's capacity for self-determination. Moreover, Origen's doctrine of the preexistence of souls has significant implications for eschatology and soteriology. He envisions salvation as a process of spiritual ascent, where the soul gradually purifies itself from the defilements of sin and ignorance, ultimately reaching union with God. The journey of the soul encompasses multiple incarnations and experiences, each contributing to its moral and spiritual growth. In this regard, Origen's theology resembles the Eastern religious concepts of karma and reincarnation, wherein the soul undergoes a series of rebirths until it achieves enlightenment or liberation. However, Origen's understanding of salvation is distinctively Christian, grounded in the redemptive work of Christ and the hope of resurrection.

Origen's teaching on the preexistence of souls was not without controversy within the early Christian tradition. His speculative cosmogony and eschatological ideas differed from mainstream Christian orthodoxy, provoking criticism from his contemporaries and later theologians. The doctrine of preexistence was deemed incompatible with the doctrine of original sin and the orthodox understanding of Christ's atonement. Furthermore, Origen's theology faced condemnation from ecclesiastical authorities, mostly in the form of censures circulated years after his death by the Fifth Ecumenical Council in 553 AD.[3] The council condemned several of Origen's teachings, including his doctrine of apokatastasis (universal restoration) and his allegorical interpretation of Scripture.

Despite these challenges, however, Origen's doctrine of the preexistence of souls continues to encourage us to reflect on the mystery of human existence and the divine purpose that animates it. His doctrine also causes us to reflect on the intricate balance between God's intervention and human free will, urging us to discover the redeeming purpose of our lives within God's plan. Origen's thoughts on the preexistence of souls combine fascinating philosophical ideas and Scripture to explore the human soul's connection to God, even if some of his views deviate from traditional Christian teachings.[4]

TERTULLIAN: TRADUCIAN MATERIALITY OF THE SOUL

According to a philosophical concept known as *traducianism*, human souls, like human bodies, are derived from the seed of the father, hence the father may transmit to his children even his own sins. Traducianist theory states that the soul is material, and the human seed passes on a portion of the parent's soul and body. Traducianists, including early Christian theologians such as Tertullian, as well as many Protestant advocates such as a number of Lutheran Churches as well as modern theologians such as Augustus H. Strong (Baptist), and Gordon Clark (Presbyterian), Lewis Sperry Chafer, Millard Erickson and Norman Geisler.[5]

Traducianism stands as a distinctive theological doctrine within Christian thought, offering an alternative perspective on

---

3. Tanner, *Decrees of the Ecumenical Councils*, 190.
4. Danielou, *Origen*, 185.
5. Erickson, *Christian Theology*, 506.

the origin and transmission of the human soul. This doctrine speculates that the human soul is generated in a manner analogous to the generation of the human body, namely through the procreative act of the parents. Essentially, traducianism holds the idea that the soul, like the body, is transmitted from parent to offspring through the natural process of reproduction.[6] Just as the physical characteristics of a child are derived from the genetic material of its parents, so too is the immaterial aspect of the human person—the soul—generated from the spiritual substance of the parental souls. This understanding implies a direct continuity between the souls of parents and children, suggesting a shared essence that is passed down through successive generations.

The theological rationale behind traducianism is multifaceted and draws upon various strands of Christian doctrine, particularly the doctrine of original sin, which holds that all humanity inherits a sinful nature because of Adam's transgression. According to traducianism, this inherent sinfulness extends not only to the physical body but also to the soul itself. Proponents of traducianism argue that if human beings are conceived in a state of original sin, then it follows that their souls, as well as their bodies, are tainted by this primordial rebellion against God, as discussed at length in the prior sections. Drawing upon passages such as Ezek 18:20, which declares, "the soul that sins shall surely die," Traducianists interpret this as evidence of the soul's complicity in Adam's sin and its subsequent corruption. Thus, each newly conceived soul inherits the stain of original sin from its progenitors, perpetuating the cycle of spiritual estrangement from God.

Traducianism offers a solution to the theological dilemma posed by the doctrine of original sin. If God is the creator of every soul *ex nihilo*, as theorized by creationism, then it would seem conflicting for him to impart a tainted nature to his creatures. This suggests that if souls come into being as part of the natural process of human procreation—meaning that a new soul is created whenever a new human life begins—then it becomes easier to comprehend how original sin could be passed down through generations.

---

6. Tertullian, "De Anima," ch. 2.

In this view, the soul isn't something separately inserted into a person after their conception but is instead inherently tied to the physical process of coming into existence. Just as genetic traits are passed down biologically from parents to children, the inclination or burden of original sin could be transmitted through this naturally generated soul. This idea fits into a biological or "living systems" framework, where the process of creating life (including the soul) is connected to inherited qualities. Essentially, it's about seeing the soul's formation as part of the same generational continuity as physical traits by means of epigenetic inheritance, which is a plausible mechanism for the transmission of original sin from one generation to the next. Just as physical traits can be inherited through genetic material, so too can spiritual predispositions and moral inclinations be passed down through the soul. This hypothetical method of a metaphysical transmission of sin suggests that there may be a profound interconnectedness of the human family rooted in the pervasive effects of Adam's disobedience on subsequent generations.

Historically, traducianism has held a place of prominence within Christian theology, particularly during the early centuries of the church. Influential theologians such as Tertullian and Augustine held to views that aligned with the core tenets of traducianism. Tertullian affirmed the notion of traducian generation in his treatise, "De Anima"—both body and soul are transmitted through the act of procreation.[7] Augustine, while not explicitly endorsing traducianism, considered similar ideas regarding the transmission of original sin through the propagation of human generations. His doctrine of seminal identity, which states that the entire human race was seminally present in Adam's loins at the time of his fall, reflects the underlying principles of traducianism. Augustine's emphasis on the corporate solidarity of humanity in Adam emphasizes the collective responsibility for sin and its consequences.

Despite its historical prominence, traducianism eventually fell out of favor within mainstream Christian theology, succeeded

---

7. Tertullian, "De Anima," ch. 2.

by competing doctrines such as creationism, which asserts that each soul is directly created by God at the moment of conception, as maintained during the Middle Ages and the Reformation era. One of the primary objections to traducianism stems from concerns regarding the transmission of moral guilt and personal responsibility. Critics argue that if souls are generated from the souls of the parents, then it would entail a form of *vicarious culpability*, whereby individuals are held accountable for sins committed by their ancestors. Such a notion runs counter to principles of justice and individual autonomy, wherein each person is judged according to their own deeds and intentions. Additionally, opponents of traducianism raise theological objections regarding the nature of the soul and its relationship to God. The idea that the soul is derived from the substance of the parental souls raises metaphysical questions about the ontological status of the soul and its dependence on external sources for its existence. This view undermines the spiritual uniqueness and divine origin of the individual soul, compromising its inherent dignity and autonomy.

One can understand why traducianism remains a subject of debate, with both supporters and critics offering perspectives on its implications. Some theologians advocate for revisiting traducianism considering recent advancements in genetics and developmental biology, both of which shape human identity. Others urge caution, pointing to unresolved theological tensions and doctrinal ambiguities, suggesting a more balanced approach that integrates insights from various theological traditions while upholding the core tenets of Christian orthodoxy. As will be discussed in subsequent chapters, however, traducianism offers a provocative view of the soul's origin and transmission, providing a compelling framework for understanding the transgenerational inheritance of sin through epigenetics.

AUGUSTINE: TRADUCIAN IMMATERIALITY OF THE SOUL

One might wonder why Augustine was not able to accept absolute traducianism without hesitation, as discussed previously. His commitment to literal solidarity in Adam implied that human composition includes material from Adam contained in the soul. Yet Augustine could not accept inherited sin as a sin of the soul and body, for he held that sin was not material substance but a corruption within the immaterial soul, which is opposed to traducianism.[8] Although creationism and preexistence theories were available to him, immaterial traducianism seemed to be the most appealing to him, yet he could never resolve how the soul could be propagated.[9] Without any resolution to this dilemma, Augustine arrived at the idea that the soul is *weighed down* by the corrupted body produced by lustful sex.[10] This idea still did not answer the problem that sin affects the body and the soul, since Augustine believed that all sinned in Adam, and all of humanity in that one man was implanted in his nature. Thus, we were in Adam because we were in his seed, and in accordance with the mysterious and powerful natural laws of heredity, those who were in his loins and were to come into this world through concupiscence or lustful desires of the flesh, were condemned with him.[11]

It is important to recognize that Augustine gives sexual desire a significant role in original sin's transmission. Augustine claims that carnal concupiscence is the essence of sin.[12] Lust becomes causally involved in the transmission of original sin, as he states that those who are born from the union of bodies are under the power of the devil, before they are reborn, because they are born through that concupiscence by which the flesh has desires opposed

---

8. Augustine, *C.Jul.imp.* 2.178.
9. Augustine, *C.Jul.imp.* 2.178; 4.104.
10. Augustine, *C.Jul.imp.* 3.44.
11. Augustine, *Pecc.Mer.* 3.7.14.
12. Augustine, *nupt.et conc.* 1.24.27.

to the spirit.[13] Sexual lust, thereby, becomes not merely a symbol of carnal concupiscence, but its cause. Augustine finally arrived at the concept of immaterial traducianism, a heuristic in which the soul, though immaterial and created by God, is "transmitted" from parent to child, carrying the inherited effects of original sin—a spiritual parallel to how certain traits or predispositions can pass across generations. Yet his reflections remain tenuous, as he never settled on a definitive theory of the soul's origin and at times seemed tempted to think that in creating Adam's soul, God had created once and for all the souls of all humanity.[14]

AQUINAS: CREATION AND TRADUCIANISM

To further Augustine's immaterial traducianism, with subsequent challenges to the theory over the centuries, notable theologians such as Thomas Aquinas also rejected absolute traducianism (implies full transmission of soul and sin nature). Aquinas chose a form of creationism on the grounds that the human soul has activities beyond the capacity of matter and the existence of these activities shows that the human soul is both immaterial and immortal, but not independent of God's causality.[15] In agreement with Aquinas, the Roman Catholic Church also rejected traducianism by stating that, every spiritual soul is created immediately by God and it is not produced by the parents, and also that it is immortal. "The rational soul is created the moment it is infused into the new organism."[16]

Aquinas, following Aristotle's embryology, taught that the rational soul is created when the antecedent principles of life have rendered the fetus an appropriate organism for rational life, though some time is required after birth before the sensory organs are sufficiently developed to assist in the functions of intelligence. The

---

13. Augustine, *C.Jul.* 4.4, 34, 79.
14. Augustine, "Nature and Origin of the Soul," 4.4.
15. Aquinas, "Union of the Soul with the Body," 1.76.
16. *Catechism of the Catholic Church*, §366.

doctrine of Aquinas concerning the human soul is essentially that when the form inherent in matter serves as the basis for actions that remain within the subject, and as such, is more particularly called the *soul*.[17]

For Aquinas, the human soul exists for the matter of the body to be organized into a living human body, which would indicate that the soul is a material form, inherent in matter. From this proposition, the soul is nothing but the specific organic structure of the matter of a living human body. Although this conception of the soul would appear to entail this materiality of the soul, Aquinas clearly believed in the immortality of the human soul. To circumvent the problem of materialism, he held that the human soul is both *subsistent* and *inherent* in matter.[18] Yet again, this is clearly a traducianistic perspective—that the soul supports itself and has its own attributes. To address this paradox, Klima explores how Aquinas refined his concept of materialism: "subsistent forms are a metaphysical argument in which material forms are received and multiplied in matter and cannot be identified with their subjects."[19] Immaterial forms are subsistent as well, Klima continues, since they are nothing but immaterial substances themselves. Such subsistent forms, according to Aquinas, are angels and God, both of which are defined first by separate metaphysical arguments.

According to Aquinas, subsistent forms do not have matter in the way material substances have matter as their component, that is, *they do not inform matter*. Because Aquinas claims that the human soul is both subsistent and inherent, the soul must be material in the sense that it is inherent in matter. Even though Aquinas does not tie the notions of subsistence or inherence of forms to their immateriality or materiality, he holds that there are inherent immaterial forms, such as the act of thinking, and he also holds that there are material forms, namely, human souls, that actually *inform our bodies*, which are, nevertheless, also subsistent.[20] Aquinas

17. Aquinas, "Union of the Soul with the Body," 1.76.
18. Aquinas, "Union of the Soul with the Body," 1.75
19. Klima, "Aquinas on the Materiality of the Human Soul," 169.
20. Klima, "Aquinas on the Materiality of the Human Soul," 170.

concludes that the soul is certainly material in the sense that it is *inherent in the matter of the human body*, simply because it has its being as that by which this matter is actualized in a human form.

In essence, Aquinas's position is that the soul is material, insofar as it is inherent in the matter of the human body, having the substantial act of being of this human, as that by which this body is actualized, organized into a living human being. Therefore, the human soul must be a subsistent being as well as an inherent being because it has *its own* activity, such as understanding or intellect. By saying that it is *its own*, Aquinas means that the intellect, the power to think, has an inherent active quality. The activity of thinking is not inherent in the soul-body composite in the way bodily activities (such as walking or perception) are, but they are inherent in the substance of the soul alone.[21] It is this inherent active quality of the soul, as described by Aquinas, that could support the action of original sin, including constitutional fault, carnal concupiscence and, as will be discussed later, epigenetic inheritance.

From Augustine to Aquinas, the inherent, active quality of the soul substance could explain the transmission of original sin as a metaphysical transaction and support the plausibility of epigenetic inheritance of the soul, thus the inheritance of original sin. It would appear, then, that original sin could have occurred as result of an environmental challenge to an otherwise perfected soul at birth, as Augustine cries:

> Who, however, will explain in words, who will at least discover in thought how the same people was in the loins of Abraham . . . but from his time up to the present time. How, then, could there be in the loins of one man so countless a multitude of human beings?[22]

The feasibility of this, however, can be observed from Paul's postulate in Rom 5:12, "Therefore, just as through one person sin entered the world, and through sin, death, and thus death came to all, inasmuch as all sinned . . . " and 1 Cor 15:22, "For just as in

---

21. Klima, "Aquinas on the Materiality of the Human Soul," 172
22. Augustine, *C.Jul.imp.* 4.104.

Adam all die, so too in Christ shall all be brought to life." These perspectives suggest that if we view the soul as having a material or substantial aspect, with an active, inherent quality enabling its independent existence, then a new category for the propagation of the soul might be considered: *neo-traducianism*. This term would propose that the soul is not only generated through human procreation but also carries properties that shape human behavior and moral disposition. Such an active quality in the soul could, in theory, allow traits like sin and guilt to be transmitted across generations. In this way, the soul's active nature might function as a mechanism through which the burden of sin and guilt is inherited, much like how epigenetic modifications are passed from parents to offspring.

The question arises, then, as to how epigenetic change is transmitted through the sperm of the male? One of the key studies that addressed this question was conducted by Pembrey et al. (2006), in which they demonstrated that sex-specific, male-line transgenerational responses exist in humans.[23] The transmissions were apparently mediated by the sex chromosomes, X and Y, which opened a new dimension in how genes are affected by environment interactions. In another study, it was found that RNA in human sperm might also affect human inheritance. RNA in the sperm and eggs of mice can transfer heritable traits, which are findings that confound our current understanding of what drives inheritance, suggesting that RNA found in human sperm may also influence genetic expression.[24] As will be discussed in later sections, our corrupted state may be present as patterns of epigenetic marks in the human epigenome because of original sin. Transgenerational epigenetic inheritance is the idea that epigenetic marks (i.e., DNA methylation and histone modifications discussed in detail in the next chapter) can be acquired on the DNA of one generation and stably passed on through the gametes (i.e., sperm and eggs) to subsequent generations.

---

23. Pembrey, "Sex-Specific, Male-Line Transgenerational Responses in Humans," 159–66.

24. Grandjean, "RNA," 1–7.

Traumatic experiences and other harmful environmental exposures can cause significant epigenetic modifications that can become imprinted in the human epigenome that persist across multiple generations.[25] It is only with the recent advances in the field of epigenetics that we can begin to hypothesize about the possible biological impacts of profound traumas, such as the banishment of Adam and Eve from the garden of Eden. The intense guilt and shame that followed this event, along with the drastic environmental changes, may have been imprinted on the epigenome of those affected. This speculation suggests that the profound psychological and environmental shifts they underwent could have led to lasting epigenetic changes in their genes, as will be described in the next chapter. These modifications may have altered their neurobiological responses to stress, anxiety, and moral awareness, effectively embedding a predisposition toward sinful behavior. Such changes could have been passed down to subsequent generations through mechanisms of epigenetic inheritance. This perspective provides a possible explanation as to how the persistent sense of shared guilt and the innate tendency to distance oneself from God is what Augustine attributes to original sin. According to this view, the inherited epigenetic changes in our genes could be a biological basis for the deep-rooted feelings of moral failure and spiritual estrangement that characterize the human experience. These traits align with Augustine's description of original sin, which suggests that humanity is born with an inherent flaw that inclines us to turn away from God. This hypothesis introduces a novel way of understanding how this formative event might have biological underpinnings, addressing the core of the human condition: original sin as the inherently disordered orientation with which we are born.

This disordered state not only inclines us toward personal sins but is, in essence, a manifestation of sinfulness. The problem lies in an inherited constitutional fault, one that has not only led humanity into a state of bondage, but one that also does not absolve us of personal responsibility. Augustine's view of original sin explains

25. Yehuda, "Public Reception of Putative Epigenetic Mechanisms," 1–7.

the *bondage of the human will* and demonstrates how the predisposition of sin could, in fact, mirror addictive behavior, in that they are both voluntary and yet beyond the immediate control of a free will. Building on a general understanding of the heritability and transmission of original sin in relation to the theology of the soul, the following chapters will investigate how the emerging field of epigenetics could help to explain the mystery of how sin might be inherited across generations while still providing evidence for change and freedom.

4

Epigenetics

A Key Scientific Window into Augustinian Theology

EPIGENETICS, IMAGINED AS A conductor orchestrating a symphony of genes within the human genome, has emerged as a revolutionary field at the intersection of genetics, molecular biology, and epistemology. The science of epigenetics explores the intricate mechanisms that regulate the activation and deactivation of genes, much like a sound engineer fine-tuning each instrument for optimal performance. Through a complex network of molecular signals and protein complexes, epigenetics examines how genes are regulated, without changing the underlying DNA sequence, resulting in changes in gene expression.

At the heart of epigenetics lies the concept of epigenetic modifications, which encompass a variety of chemical alterations to the structure of DNA, particularly methylations, which are the primary source of alterations discussed in this book. These modifications serve as *marks* that determine the accessibility of genes to the cellular mechanisms responsible for transcription and translation—the processes by which genetic information is converted into functional proteins.

One of the most well-studied epigenetic modifications is DNA methylation, wherein methyl groups are added to specific cytosine residues within the DNA sequence. This modification typically occurs at CpG dinucleotides, regions of DNA where a cytosine nucleotide is followed by a guanine nucleotide. DNA methylation plays a pivotal role in gene regulation, serving as a mechanism for silencing gene expression and maintaining genomic stability. Histone modifications, also noted periodically in this book, are another aspect of epigenetic regulation that involve chemical alterations to the histone proteins around which DNA is wrapped. These modifications, which include acetylation, methylation, phosphorylation, and ubiquitination, influence the structure and function of chromatin—the complex of DNA and histones that constitutes the chromosome. By altering the degree of chromatin compaction, histone modifications regulate the accessibility of genes to transcription factors and RNA polymerase, thereby modulating gene expression.

Epigenetic regulation also encompasses non-coding RNAs, such as microRNAs (miRNAs) and long non-coding RNAs (lncRNAs), which play key roles in post-transcriptional gene regulation. These small RNA molecules can bind to messenger RNAs (mRNAs) and either promote their degradation or inhibit their translation into proteins, thereby exerting fine-tuned control over gene expression.

The nature of epigenetic regulation is featured by its responsiveness to environmental cues and developmental signals. External factors such as diet, stress, toxins, and social interactions can influence epigenetic modifications, leading to changes in gene expression that may have profound implications for health and disease. Moreover, epigenetic changes acquired during early development or in response to environmental stimuli can be transmitted across generations—a phenomenon known as epigenetic inheritance. The implications of epigenetics extend far beyond the science of molecular biology—it extends into fields as diverse as medicine, psychology, anthropology, philosophy, and with this work, theology. In medicine, epigenetics has revolutionized our understanding of disease etiology and pathogenesis, revealing the intricate chemistry between genetic predisposition and environmental factors in shaping health outcomes. Epigenetic biomarkers hold promise for early detection, prognosis, and personalized treatment of various diseases, including cancer, neurodegenerative disorders, and metabolic syndromes. In psychology and psychiatry, epigenetics provides insight into the molecular mechanisms that shape behavioral traits and mental disorders. Recent studies link epigenetic dysregulation to conditions such as depression, anxiety, schizophrenia, and addiction, illustrating how genetic vulnerability and environmental stressors interact in the development of these disorders.

Understanding the epigenetic foundations of behavior has important implications for therapeutic and preventive approaches, including epigenetic editing. Beyond mental health, epigenetics can also explain how environmental adaptation and phenotypic plasticity drive species diversity by enabling organisms to respond swiftly to changing conditions. From a theological perspective, epigenetics challenges traditional notions of genetic determinism—that genes alone dictate an organism's phenotype and destiny. Instead, epigenetics emphasizes the interaction between genes and the environment, emphasizing the plasticity and adaptability of living systems. This view has a more systematic approach to understanding biology, wherein gene expression is driven by environmental factors that converge to shape the complexity of life.

Considering this, epigenetics raises profound questions about the nature of human identity, autonomy, responsibility, and more recently, affordances, based on the implications for epigenetic inheritance. If epigenetic modifications acquired during an individual's lifetime can be passed on to future generations, then to what extent are individuals shaped by their ancestors' experiences and environmental exposures? How do we reconcile notions of personal agency, free-will and accountability with the recognition of epigenetic influences on behavior, health, and spiritual outcomes? These inquiries intersect with bioethical considerations regarding the use of epigenetic information in various domains, including healthcare, forensics, and reproductive medicine. The ability to edit epigenetic changes in DNA raises bioethical dilemmas regarding privacy, consent, equity, and justice, as well as questions about the potential for unintended consequences and unforeseen risks.[1] As we navigate the ethical terrain of epigenetics, we must remain vigilant in safeguarding individual rights, promoting transparency and accountability, and equitable access to emerging technologies and therapies. To be certain, the recent science of epigenetics represents a paradigm-shift in understanding the hard-wired nature of our genes, to a reconsideration of the fundamental assumptions about life, identity, and inheritance. As we continue to unravel the mysteries of the epigenome, which is the entirety of epigenetic marks on the human genome, we are confronted with profound questions about the nature of humanity. With these profound concepts underlying epigenetics, it can now be possible to gain further clarity as to why some individuals are more susceptible to addictive behaviors while others are not, or how traumatic experiences, such as combat, can influence the likelihood of developing addictive tendencies.

Epigenetics offers insight into the environmental factors that drive gene expression and can often leave imprints on our genes, influencing their activity throughout our lives and potentially impacting future generations. To comprehend the epigenetic pathways leading to addiction, the subsequent section will discuss the biological mechanisms of epigenetics, along with the concept of transgenerational epigenetic inheritance. Following the discussion of Augustine's theology of original sin in the previous section, coupled with a general understanding of the science of epigenetics, we will then be equipped to unveil Augustine's concept of the invisible and intangible power of the propagation of original sin:

> Some sort of invisible and intangible power is in the secrets of nature where the natural laws of propagation are concealed, and on account of this power as many as were going to be able to be begotten from that one man by the succession of generations are certainly not untruthfully said to have been in the loins of the father.

1. Wing, "On Gene Editing," 293–314.

They were there . . . though unknowingly and unwillingly, because they did not yet exist as persons who could have known and willed this.[2]

From this seemingly prophetic quote, repeated here again, the transmission of original sin through successive generations finds a parallel in the emerging science of genetics, particularly epigenetic inheritance. Augustine's theological elaboration proposes that Adam's sin was not merely his alone but imputed to all his descendants, implicating humanity in a shared sinful nature inherited from our primal ancestors, Adam and Eve. From a Pauline standpoint, the repercussions of Adam's transgression reverberate through humanity, summarized in the pivotal statement, "Therefore, just as through one man sin entered into the world . . . " (Rom 5:12). This direct linkage between Adam's sin and the universal state of fallenness is emphasized further in subsequent verses, which outline the far-reaching effects of sin and the transformative grace found in Christ's righteousness. The moral and biological ramifications of original sin extend to the very core of human existence—the mind, soul, and body. The recent advancements in epigenetics shed light on how environment can induce changes in gene function throughout an individual's lifespan. These epigenetic modifications are influenced by personal choices and environmental circumstances, or affordances, and have the potential to be biologically transmitted not only to offspring but also to subsequent generations.

EPIGENETICS: THE SCIENCE OF GENE REGULATION AND EXPRESSION

Interdisciplinary studies, especially those involving biology, often encounter a significant hurdle in the form of scientific terminology. This challenge becomes particularly pronounced when delving into fields like epigenetics, which is deeply connected to complex biological concepts and processes. To conceptualize this complex

---

2. Augustine, *C.Jul.imp.* 4.104.

terrain effectively and facilitate a meaningful dialogue between science and theology on epigenetics, it is essential to provide readers with a foundational understanding of cell and molecular biology. At the heart of epigenetics lies a multitude of biological terms and concepts such as chromatin remodeling, DNA methylation, histone modifications, and non-coding RNAs, all of which form the backbone of epigenetic mechanisms. Without an understanding of these fundamental biological principles, engaging in a comprehensive discussion on epigenetics can be challenging—yet it is a challenge worth accepting as the science–theology dialogue continues.

Cell biology, the foundation upon which molecular biology is built, is a study of the structure and function of cells, the basic units of life, where the molecular processes that drive gene expression occur. Cellular anatomy and physiology form the foundation for understanding epigenetic regulation. From the nucleus, where genetic material like DNA and RNA are located, to the cytoplasm, where protein synthesis occurs, each aspect plays a crucial role in this process. The essence of molecular biology is the explanation of DNA structure and function, where the blueprint of life is encoded within the double helix. The processes of transcription and translation, in which genetic information from DNA is transcribed into RNA and then translated into proteins, demonstrate the connection between genotype (genes) and phenotype (proteins). Furthermore, molecular biology studies the relationship of various biomolecules, including proteins, nucleic acids, carbohydrates and lipids, each contributing to the various cellular functions.

Familiarizing oneself with these fundamental molecular processes and genetic regulation lays the foundation for understanding epigenetic mechanisms. As a note of encouragement and voice of support for both scientists and theologians, interdisciplinary studies pose unique challenges, particularly in fields of study like biology, as it operates within its own specialized language. In deepening our understanding of the biological principles described here, our ability to appreciate connections between scientific discovery and theological inquiry in this intriguing area of study also deepens.

Using computers as an analogy, if genetics is the hardware of DNA, then epigenetics is the software. Epigenetics and genetics work together to provide an integrated mechanism of gene expression. Epigenetics is concerned with how the genes function, how genes are read by cells to produce a specific protein. Epigenetics is essentially understanding how a gene is *tuned* by turning genes "on" and "off." C. H. Waddington first coined the term *epigenetics* to describe the mechanisms involved in programming identical genes differently in different organs during embryogenesis.[3] A more current definition describes epigenetics as *heritable changes in gene expression that operate outside of changes in the DNA sequence itself.*[4] In the mammalian system, including humans, the intricate network of cells comprise numerous specialized types—210, to be precise. These cells, except for red blood cells, contain an inner organelle known as the nucleus, which houses the DNA molecule, or deoxyribonucleic acid. This DNA molecule serves as the master blueprint, directing all cellular activities.

Human DNA is an extraordinarily complex molecule, comprised of two complementary chains of approximately three billion nucleotides or bases. Analogous to the coding languages used to operate computers, the language of DNA relies on four fundamental types of nucleotides: adenine (A), cytosine (C), guanine (G), and thymine (T). The specific sequence of these nucleotides, referred to as the *genetic code*, determines the instructional units of life known as *genes*. All of the approximately 23,000 genes within a strand of DNA constitute what is known as the *genome*. This genome serves as the source of genetic information that defines an individual's traits and characteristics. The sequence of nucleotides in a gene serves as a blueprint for the synthesis of messenger RNA (mRNA). This mRNA molecule acts as an intermediary messenger, conveying the genetic instructions from the DNA into the cytoplasm where protein synthesis occurs. The

3. Waddington, "Canalization of Development," 1654–55.
4. Robertson, "DNA Methylation in Health and Disease," 11–19.

process of protein synthesis comprises two distinct stages: transcription and translation. Transcription involves the conversion of DNA-encoded genetic information into mRNA molecules, while translation involves the subsequent conversion of mRNA into functional proteins. Through the intricate processes of transcription and translation, genes are expressed, producing an organism's traits through the synthesis of specific proteins. These proteins, in turn, serve as the molecular building blocks responsible for a vast number of biological functions, ranging from eye color and liver function to susceptibility to diseases such as depression and diabetes.

It is important to note that the human genome comprises an extensive repertoire of approximately 23,000 genes, each contributing in varying degrees to the manifestation of an individual's traits and characteristics. Thus, gene expression forms the foundation of biological diversity and individual variation observed within the human population.

The composition of DNA is in a double helix so that the bases form a ladder-like construct with each step as a base pair: A-T and C-G. Within the 6.4 billion base pairs that compose the human genome, most of the genome is composed of *non-coding DNA* and *repeat sequences of DNA* (also considered non-coding). What is a most extraordinary feature of human DNA is the fractional content of the protein-coding regions (~23,000 genes, with approximately 1,500 base pairs in a protein-coding region of a gene), which are less than 2 percent of the entire genome. The question naturally arises: what is the function of all the extra DNA? Biologists had regarded this extra non-coding and repeat regions as *junk* DNA; however, it has recently been determined that these regions are anything but junk. With the entire sequence of the human genome being completed over twenty years ago, the science is still in its very early stages of understanding how the genes function. It has been a relatively recent finding, however, that the non-coding and repeat regions of DNA play essential roles in gene expression. The DNA molecule is composed of regions that can be transposed, inverted, spliced, or edited. Essentially, DNA gives the instructions for all of the proteins to be produced inside the cell for the rest of the body. For example, the *MLANA* gene in DNA is present in all the 210 types of cells in the body, but is only "expressed" by the genes in skin cells to produce the protein melanin.[5] Epigenetics controls *how* and *where* the *MLANA* gene is expressed and in which cell this occurs. Thus, it is by epigenetics that a cell's specialization (e.g., skin cells, blood cell, hair cell, liver cells, etc.) is determined. A fetus develops into a baby based on the genes that are expressed throughout the stages of development (active). The genes that are not necessary during a specific stage of development are silenced (dormant) due to *environmental* stimuli that can cause genes to be turned "on" or "off." The process by which genes are turned "on" or "off" is called *epigenetic gene regulation*.[6]

Communication molecules from outside of the cell, such as growth factors, have the capability of notifying the cell when to turn a growth gene "on" or "off." Each of these environmental factors can cause chemical modifications (such as communications to the cells to grow) to the genes that will turn those genes "on" or "off" either immediately or over several generations.[7] Additionally, in certain diseases such as cancer or even addictive behavior, various genes can be switched from a normal or healthy state to an abnormal or diseased state or silenced altogether.

With every gene having "on" and "off" switches, some switches remain on or remain off for extended periods of time, some for an entire lifetime. Genes that are permanently on are mostly structural genes. The genes that are the most responsive to change include behavioral genes, which can be altered based on environmental conditions, such as social interactions, nutrition, climate, etc. The critical finding in the description of epigenetics is that genes are controlled from the inside and outside of the body. Different combinations of genes that are turned "on" or "off" make each human being unique. Some of these epigenetic changes can

---

5. Edqvist, "Expression of Human Skin-Specific Genes," 129–41.
6. Yan, "Epigenetics in the Vascular Endothelium," 916–26.
7. Szyf, "DNA Methylation," 49–57.

be inherited and even be transgenerational, as will be elaborated upon in later chapters.

Interestingly, a key difference between the *genetic* code and the *epigenetic* code is that many epigenetic changes are reversible.[8] With changes in the expression of the 23,000 genes within the human genome, the number of different combinations of genes being turned "on" or "off" could be astronomical. Conceptually, then, if it were possible to map every single cause and effect of the different combinations of genes being expressed, it could be possible to reverse an abnormal/diseased gene's state to a normal/healthy state. The possibilities being explored today hold extraordinary potential for treating diseases like cancer, Alzheimer's, and MS, as well as preventing relapses in addictive behavior.

DNA METHYLATION AND THE METHYLSCAPE

The primary epigenetic mechanism involved in gene regulation is *DNA methylation*, which is a chemical modification that alters gene expression and allow cells to respond and adapt in response to their environments. Epigenetic markers, such as DNA methylation, allow for the transmission of gene expression from one cell to its daughter cells and plays a significant role in gene-environment interactions, particularly in psychiatric disorders.[9] DNA methylation, which is considered the *epigenetic code*, is a stable epigenetic mark that has important roles in mammalian development, differentiation, and maintenance of cellular identity through the control of gene expression. DNA methylation occurs on the cytosine nucleotide, repressing gene activity. Conversely, when specific regions of a gene lack methylation, the gene can be expressed. DNA is wrapped up in the cell around histone molecules. Several molecules can also be attached to the histone tails resulting in an alteration of the activity of the DNA.[10] While epigenetic modifica-

8. Ramchandani, "DNA Methylation Is a Reversible Biological Signal," 6107–12.

9. Kubota, "Epigenetic Understanding of Gene-Environment," 1.

10. Qiu, "Epigenetics," 143–48.

tions to histones will be discussed briefly in this book, they will take center stage in a future work when addressing other social issues, such as trauma.

DNA methylation is a covalent modification of DNA induced by the addition of a methyl group to cytosines in dinucleotide CpG sequences that results in *gene silencing*.[11] Cytosine methylation is mediated by a family of DNA methyltransferase enzymes, namely DNMT1, DNMT3A, and DNMT3B. These enzymes are responsible for attaching methyl groups to the cytosines of DNA, a process that can be likened to adding punctuation marks to a sentence. Environmental factors such as stress[12] and toxins[13] have been shown to alter DNA methylation at specific genes in germ cells (sperm or ova) that can be transmitted across multiple generations. More importantly, DNA methylation is involved in *genomic imprinting*, a process that allows for the selective expression of only one parental gene in the offspring.[14] At imprinted regions (*loci*), gene silencing is mediated by hypermethylation of DNA regions and is reproduced in the daughter cells.[15] Together, these studies demonstrate that DNA methylation is a mechanism for the transmission and maintenance of epigenetic alterations in response to the environment. Over the past forty years, changes in DNA methylation patterns, referred to as the *methylscape*, have been observed in many human diseases, such as cancer, behavioral disorders, and particularly addictive behavior.[16]

An example of how the methylscape is affected by the environment occurred in the Netherlands after WWII during the Dutch Hunger Winter of 1944 to 1945.[17] The study investigated

11. Szyf, "DNA Methylation," 49–57.
12. Franklin et al., "Epigenetic Transmission," 408–15.
13. Guerrero-Bosagna et al., "Epigenetic Transgenerational," e13100.
14. Sha, "Mechanistic View of Genomic Imprinting," 197–216.
15. Skinner, "Role of Epigenetics in Developmental Biology and Transgenerational Inheritance," 51–55.
16. Sina et al., "Epigenetically Reprogrammed Methylation Landscape," 4915.
17. Heijmans et al., "Persistent Epigenetic Differences Associated with

how the body could remember (over decades) the environment it was exposed to in the womb. Bas Heijmans, a geneticist at Leiden University Medical Center in the Netherlands, examined blood samples obtained from individuals whose mothers were pregnant with them during the Dutch Winter of 1944 to 1945. Blood samples were also analyzed from their siblings who were born either before the six-month famine or after. This cohort of individuals whose mothers were pregnant during the famine had been shown to have higher rates of obesity, dyslipidemia, diabetes, and schizophrenia. The study evaluated the methylation patterns of the insulin growth factor gene IGF2, which is one of the best characterized epigenetically regulated genes and is a key factor in human growth and development. The study concluded that the Dutch Hunger Winter *silenced* or altered the methylation patterns of certain genes in unborn children—and that the genes have stayed silent ever since. These results were the first to support the hypothesis that early-life environmental conditions can cause epigenetic changes in humans that persist throughout life.

BEHAVIORAL EPIGENETIC INHERITANCE

Epigenetic transmission can also occur in the body (somatic) cells, independently of the germ cells, and can also involve behavioral or social transmission.[18] During behavioral/social transmission, epigenetic changes occur when environmental factors bring about epigenetic modifications that persist in the environment. In a landmark behavioral study conducted in mice, it has been demonstrated that in early life, variations in maternal licking and grooming behavior can alter epigenetic marks throughout the genome and determine the level of care administered to the subsequent generation.[19] As shown in Figure 2, daughters receiving greater levels of maternal care will impart increased levels of

---

Prenatal Exposure," 17046–49.

    18. Bohacek, "Epigenetic Inheritance of Disease and Disease Risk," 220–36.

    19. McGowan, "Broad Epigenetic Signature of Maternal Care in the Brain of Adult Rats," e14739.

maternal care on their offspring. In cross-nurturing studies, pups from high licking and grooming mothers transferred to low licking and grooming and anxious mothers not only demonstrated decreased maternal behavior towards their subsequent offspring, but also displayed epigenetic modifications like those naturally born to low licking, grooming, and anxious mothers.[20] These results concluded that this effect is not permanent and must be instated or reinstated for each new generation. These studies demonstrate that training human mothers to provide good maternal care may have positive transgenerational effects.[21]

Similarly, Mychasiuk et al. also demonstrated that paternal stress also had negative effects on behaviors and increased DNA methylation in the hippocampus of their offsprings.[22] In contrast, enrichment of the environment of male Long Evans rats with toys, multiple levels of exploration, and several cage mates for 28 days before mating with control female rats had a positive impact on the exploratory behaviors of the males' offsprings and on DNA methylation in their hippocampi and frontal cortices. These observations suggest that promoting resilience by enriching the environment may reduce the abnormal behavior due to stress (such as addictive behavior) in the offsprings of parents reared in enriched environments.

TRANSGENERATIONAL PROPAGATION BY IMPRINTED GENES

Gene imprinting, also known as *genomic imprinting*, is an epigenetic phenomenon where certain genes are methylated that affect their expression based on the parent of origin. These epigenetic changes can be inherited over several generations. The process of gene imprinting involves DNA methylation of certain genes during gametogenesis (sperm and egg formation). The methyl groups

20. Weaver, "Epigenetic Programming by Maternal Behavior," 22–28.
21. Weaver, "Epigenetic Programming by Maternal Behavior," 847–54.
22. Mychasiuk, "Parental Enrichment and Offspring Development," 294–98.

attached to DNA are maintained throughout development and are responsible for guiding the differential expression of imprinted genes based on whether they were inherited from the mother or the father. When gametes (sperm and eggs) are formed, imprinted genes are tagged with methyl groups in a sex-specific manner. For example, if a gene is imprinted to be more active when inherited from the father, it will receive certain epigenetic modifications (specifically, DNA methylation) that promote its expression, while the same gene inherited from the mother might have the opposite modifications, leading to reduced expression.[23] The genes that methylated are then maintained during fertilization and throughout subsequent development. When the fertilized egg divides and forms a multicellular embryo, the methyl groups guide the expression of imprinted genes according to their parental origin. It is important to note that other epigenetic mechanisms, such as histone acetylation and non-coding RNAs, involved in gene imprinting can vary as understanding of this phenomenon continues to evolve.

Imprinted genes play crucial roles in growth, development, and metabolism, yet disruptions in gene imprinting can lead to various developmental disorders, such as the loss of insulin and insulin-like growth factor1 (IGF1) receptors in cells.[24] The loss of these receptors induces a major decrease in expression of multiple imprinted genes, regardless of whether they would be maternally or paternally expressed. This deregulation is associated with changes in the methylscape and provides an alternative pathway by which insulin and IGF1 receptors may regulate growth and metabolism during early development and thereafter. Understanding that gene imprinting can result in epigenetic modifications that are established during gametogenesis, the modifications can be carried forward across generations as they contribute to the complex regulation of gene expression in a parent-of-origin-specific manner. These findings provide a plausible mechanism for

23. Von Meyenn et al., "Forget the Parents," 1248–51.

24. Boucher et al., "Insulin and Insulin-like Growth Factor-1 Receptors," 14512–517.

transgenerational epigenetic inheritance, particularly in light of the recent discovery of the ZFP57 protein, the master regulator of genomic imprinting.

A recent study by Jiang et al. has found that ZFP57, a KRAB zinc finger protein, controls imprinted expression of targeted imprinted genes primarily through maintaining differential DNA methylation at the imprinting control regions (ICRs) of the genome.[25] This study demonstrated that ZFP57 is a master regulator of genomic imprinting since it maintains DNA methylation at most known imprinted regions and regulates allelic expression of the corresponding imprinted genes. The study concluded that ZFP57-dependent regulation of parent-of-origin-dependent expression of the imprinted genes includes a mechanistic link between DNA methylation imprint and monoallelic expression of the imprinted genes. In essence, the authors of the study determined that ZFP57 regulates allelic expression of clustered imprinted genes through maintenance of DNA methylation at the imprinting control regions.

Understanding the mechanism of genomic imprinting offers valuable insights into inherited diseases as well as behavioral disorders, and even allows speculation as to how original sin might have altered the human methylscape by leaving an imprint on the human genome, which is a topic we will explore in a later chapter. Yet, with regard to genomic imprinting, questions arise such as, why are some epigenetic modifications permanent, or imprinted, while some may be transient? What causes a genetic change versus an epigenetic change, realizing that genetic changes alter the sequence of DNA that result in mutations, whereas epigenetic changes modify the expression or methylscape of the gene without altering the underlying DNA sequence? Answering these key questions will involve a detailed exploration of the methylscape: how it is established, how it responds to environmental stimuli, and how it undergoes continual reshaping throughout an individual's life. This examination lays the foundation for understanding the

---

25. Jiang, "ZFP57 Dictates Allelic Expression Switch of Target Imprinted Genes," e2005377118.

potential transgenerational inheritance of epigenetic imprints and their theological implications.

TRANSFORMATIONAL CHANGEABILITY OF THE METHYLSCAPE

In their groundbreaking work, Champagne and colleagues discovered a causal role for epigenetic mechanisms in behavioral regulation.[26] They demonstrated that variations in maternal care led to reversible changes in DNA methylation within the glucocorticoid receptor promoter, resulting in significant effects on reactivity to stress later in life. This seminal discovery set the stage for the investigation of new sites of methylation that are both behaviorally induced and specific for a gene, such as the *BDNF* gene (discussed in detail in a later chapter). By the mechanism of the ten-eleven translocation (TET) enzymes, methyl groups from the 5-methyl cytosines are removed, which means that the gene is now turned to the "on" switch. This could result in the gene being expressed, if it was originally silenced. Expression of a silenced gene results in the production of protein that was not being produced, indicating that DNA methylation can be reversible.[27] Essentially, genes that could be related to a specific disease or behavior, such as addiction, but with the expression of these genes being variable, could be expressed or silenced. Recent studies by Kaplan et al. suggest that the TET enzymes not only contribute to memory formation and trauma processing, but they also play a role in addiction-related changes in the brain, where alterations in DNA methylation patterns affect addiction susceptibility and behavior.[28] Kaplan's research observed that TET1, a key enzyme included in the TET enzymes, influences synaptic plasticity in brain regions linked to reward and learning. Since addiction involves the formation of intense, long-term memories of drug use and associated cues,

---

26. Champagne, "Transgenerational Effects of Social Environment," 266.
27. Rasmussen et al., "Role of TET Enzymes," 733–50.
28. Kaplan et al., "DNA Epigenetics in Addiction Susceptibility," 1–5.

TET1-mediated DNA demethylation is thought to facilitate these epigenetic changes. These processes involve interaction with the JMJD3 protein and other histone-modifying enzymes, described by Heller et al., as they enhance vulnerability to addictive behaviors by altering neural pathways related to habit formation and reward processing.[29] The interaction between JMJD3 and TET1 highlights a critical axis in epigenetic regulation, with profound implications for understanding addiction, trauma, and transgenerational inheritance. TET enzymes, particularly TET1, mediate DNA demethylation processes influenced by chronic drug exposure, reinforcing addictive behaviors by altering neural pathways. Similarly, JMJD3 contributes to resetting epigenetic marks in germline cells, supporting the persistence of heritable epigenetic changes. Together, these enzymes coordinate a changeable epigenetic methylscape that supports susceptibility to addiction and its generational transmission.

Within this methylscape, Augustine's depiction of original sin finds a thought-provoking parallel through theological speculation. Just as Augustine conceived original sin as an inherited disposition, modern epigenetics reveals how life experiences, such as trauma and addiction, imprint the genome, shaping behaviors and susceptibilities passed to subsequent generations. The original sin Augustine described, can be creatively reflected upon in light of modern understandings of stress-induced epigenetic reprogramming, where JMJD3 and TET1 play pivotal roles. These findings propose a novel integration: the doctrine of original sin as a theological foundation for understanding the biological inheritance of human tendencies and behaviors. This perspective reframes inherited sin as an epigenetic continuum, where environmental and emotional influences create molecular imprints that shape dispositions passed through generations.

While epigenetic modifications were once thought to be fully erased in germ cells, current research reveals that this reset is incomplete. Acquired epigenetic changes, including those linked to stress or behavior, can persist and be transmitted to descendants,

---

29. Heller et al., "Neuroepigenetics of Addiction," 1140–49.

offering a biological mechanism for the inheritance of traits, including susceptibilities, that support the theological concept of a heritable human condition. This phenomenon is known as *transgenerational epigenetic inheritance*.[30] This mechanism can then allow for the adult onset of diseases and acquired traits because the somatic cells that are derived from this germline will acquire the modified epigenome.[31] While this field is in its infancy, several studies have demonstrated mechanisms of transgenerational epigenetic inheritance. In the example introduced earlier, when exposed to high levels of environmental toxins such as vinclozolin (an endocrine disrupter found in agriculture), the DNA methylation pattern of the sperm three generations removed from the initial exposure (F3 males) is altered at specific regions of a gene.[32] These results are not limited to one environmental toxin as alterations in the DNA methylation pattern of F3 generation sperm, but were also observed following exposure to pesticides, plastics, and dioxin.[33] These exemplary studies demonstrate that exposure to environmental toxins during gestation induces permanent epigenetic changes in the germline that transmits adult-onset diseases to future generations in the absence of any subsequent exposure.

Another relevant example involves a model that was recently developed to examine early life stress by unpredictable maternal separation and maternal stress. This manipulation severely affected behavior across multiple generations.[34] Behavioral changes caused by maternal separation and stress were linked to long-term molecular alterations in stress pathways and serotonergic signaling. These changes were transmitted across F1–F3 generations. Analysis of sperm revealed altered DNA methylation patterns, which

30. Vassoler et al., "Mechanisms of Transgenerational Inheritance," 198–206.

31. Skinner et al., "Epigenetic Transgenerational Actions," 214–22.

32. Guerrero-Bosagna et al., "Epigenetic Transgenerational Actions," e13100.

33. Manikkam et al., "Dioxin (TCDD) Induces Epigenetic Transgenerational Inheritance," e46249.

34. Franklin, "Epigenetic Transmission," 408–15.

correlated with brain changes in stressed males from subsequent generations. Stress during critical developmental periods (e.g., prenatal or early life) is not the sole context for transgenerational stress transmission. In a chronic social-defeat paradigm with adult male rats, their offspring—both male and female—exhibited heightened anxiety and depressive-like behaviors, demonstrating cross-generational impacts of adult stress.[35] However, it appears that at least some of the sibling's actions were transmitted behaviorally because only the forced swim test phenotype persists following in vitro fertilization.

Parental diet is another factor that has been shown to have transgenerational behavioral and physiological effects. Following a paternal high-fat diet in mice, myriad negative effects on offspring development have been observed.[36] Additionally, it was shown in mice that diminished reproductive and gamete function were transmitted through the first generational paternal line to both sexes of the second generation and the maternal line to second-generation males.[37] The results from these studies suggest that there are changes that have occurred in the epigenome of the germ cells. Thus, in the same way as parents can pass on genetic characteristics to their children, they would also be able to pass on all kinds of "acquired" epigenetic characteristics. The results also suggest that if these dietary changes resulted in epigenetic changes, one could infer that those powerful life-threatening experiences—such as survival from starvation, torture, or persecution—may cause epigenetic changes as well. Such environmental conditions would leave an imprint on the genes within the germ cells of both parents and pass along new traits in a single generation or to multiple generations, a process referred to as transgenerational trauma.[38]

In terms of drugs of abuse, as will be discussed in subsequent sections, there are many studies that have examined the behavioral

35. Dietz et al., "Paternal Transmission," 408–14.
36. Binder et al., "Parental Diet-Induced Obesity," 804–12.
37. Fullston et al., "Diet-Induced Paternal Obesity," 1391–400.
38. Kellermann, "Epigenetic Transmission of Holocaust," 33–39.

and physiological consequences of pre-conception drug exposure.[39] One of the first studies to show a potential mechanism of transmission demonstrated that the expression of DNA methyltransferase-1 (DNMT-1), the enzyme that facilitates the attachment of methyl groups to DNA, was decreased in the seminiferous tubules of the testes following cocaine self-administration.[40] A reduction in DNMT-1, which also plays a major role in maintaining methyl groups on imprinted genes in germ cells, could also be a potential mechanism of transgenerational epigenetic inheritance.

Although it is well documented that several genetic and environmental factors may lead to the development of addiction, the genetic basis for family history as a risk factor for developing addiction must be interpreted cautiously.[41] Addiction is influenced by genetic and environmental factors, even though the exact contribution of each of these factors is complicated and not easily distinguished. Current epigenome-wide studies are seeking to identify the set of genes and epigenetic mechanisms that are involved in addiction.

NEO-TRADUCIANISM—A PLAUSIBLE MECHANISM FOR EPIGENETIC INHERITANCE

Investigations of the known mechanisms involved in the epigenetic inheritance of traits and/or disorders because of exposure to toxins, stress, and diet are likely to be comparable irrespective of the environmental stimuli. As a toxicologist, I have observed that ancestral environmental exposures to non-mutagenic agents can exert effects in unexposed descendants.[42] Intergenerational and transgenerational inheritance provide a framework for understanding how molecular changes in the germline are influenced by environmental exposures and passed through generations.

---

39. Vassoler et al., "Impact of Exposure to Addictive Drugs," 269–75.

40. Vassoler et al., "Mechanisms of Transgenerational Inheritance," 198–206.

41. Goldman et al., "Genetics of Addictions," 521–32.

42. Laney et al., "What Is Transgenerational Epigenetic Inheritance?"

Intergenerational inheritance occurs when a non-mutagenic agent affects the reproductive cells of an individual, directly exposing their children. In pregnant females, this extends to the reproductive cells of their offspring, resulting in exposure to their grandchildren, a phenomenon classified as multigenerational inheritance. Transgenerational inheritance, however, involves changes that manifest only in the great-grandchildren or later generations, where molecular alterations in the germline persist due to imprinting but remain susceptible to environmental factors over time.

The biological model of transgenerational epigenetic inheritance correlates with the theological doctrine of traducianism, which states that the soul, like the body, is transmitted through natural generation. According to this view, human propagation would include both material and immaterial properties, suggesting that an individual's soul is derived from the soul of one or both parents. Herein is the mystery as to how inherited physical and spiritual characteristics may interrelate, suggesting a perspective on the connection of biology and theology. If this speculative proposition has credence—that is, if Adam's soul was directly created by God, while Eve's substance, both material and immaterial, was taken from Adam—then the contemporary version of traducianism discussed previously, could imply that Adam's original sin could be passed down through his descendants via transgenerational epigenetic inheritance, a concept we might call *neo-traducianism*. In this view, the epigenetic "imprint" of original sin could result either from Adam's own sinful actions, which induce inheritable changes in gene expression, or from God's response to Adam's sin, which establishes a theological and moral ordering reflected in the human epigenome; in either case, certain genes would carry marks of this inheritance, present on one parental copy and effectively silenced, creating a molecular parallel to the theological notion of inherited guilt. Studies in embryogenesis have demonstrated that silencing marks that occur in the egg and/or sperm are retained in the embryo.[43] Further, it has also been shown that approximately

---

43. Boucher et al., "Insulin and Insulin-Like Growth Factor-1 Receptors,"

100 genes are known to be imprinted and are normally selectively expressed, depending on the parent from whom the gene is inherited. Epigenetic changes that are imprinted can also be permanent and be transmitted from parents to their offspring over many generations.

Although this speculative concept of neo-traducianism has yet to be introduced into the scientific or theological communities, it is worth considering this mechanism as a bridge from science to theology, that imprinted epigenetic modifications of Adam's epigenome by original sin could be within the human genome by the mechanisms described for transgenerational epigenetic inheritance. Only centuries later could this even be hypothesized. Again, recognizing that this may be a speculative proposition, it is still worth responding to Augustine's question, *who will at least discover in thought . . .*

> Who, however, will explain in words, who will at least discover in thought how the same people was in the loins of Abraham, not only from his time up to the time mentioned in the Letter to the Hebrews, but from his time up to the present time and from now to the end of the world, as long as children of Israel are born, generation after generation? How, then, could there be in the loins of one man so countless a multitude of human beings?[44]

Only by understanding the rudimentary elements of epigenetics, as described, has it been possible to speculate how original sin could have been propagated as Augustine argued:

> that sin has been passed down intact to every member of the human race, and that sin cannot be alleviated by any act of will because it is inherent in the human being from birth.[45]

---

145 12–517.
44. Augustine, *C.Jul.imp.* 4.104.
45. Augustine, *Civ.Dei.* 14.7.

As Augustine reiterated, original sin is the result of the will of Adam,

> in whom we all originally existed when he damaged our common nature by his evil will.[46]

Understanding that human beings since Adam inherit a propensity towards sin, this would affirm that humans *are born in sin* and *in sin did our mother conceive us*, according to Ps 51:5. Additionally, Paul added that we are *by nature children of wrath* (Eph 2:3), and because of the fall of Adam and Eve, *all sinned in Adam* (Rom 5:12). It is important to note, however, that while Augustine emphasizes the intact transmission of sin, the argument presented here suggests that epigenetic inheritance offers a conceptual, metaphorical parallel—dynamic and reversible in its expression—without compromising Augustine's theological claim of universal guilt.

In his current and highly relevant research paper, "Developmental Origins of Disruptive Behavior Problems: the 'Original Sin' Hypothesis, Epigenetics and Their Consequences for Prevention" Richard Tremblay explores the developmental origins of disruptive behavior (DB) problems, including aggression, opposition-defiance, rule-breaking, and stealing-vandalism.[47] His research emphasizes the importance of understanding these behaviors that are not merely products of environmental factors or learned behavior, as traditionally thought, but as universally present in early childhood, diminishing with socialization over time. For children who fail to learn socially accepted behaviors, a "disease" status is applied, highlighting the complexity of genetic, environmental, and epigenetic influences that contribute to this failure. Interestingly, the *original sin* hypothesis aligns with Tremblay's perspective, in that human behaviors are an inherited condition of sinfulness passed down from Adam. From this hypothesis, Tremblay explains that there is a natural propensity toward rebellion against God's law, echoed in the universal presence of disruptive

---

46. Augustine, *C.Jul.imp.* 4.90.

47. Tremblay, "Developmental Origins of Disruptive Behavior Problems," 371.

behaviors in early childhood. As humans are born into sin, Tremblay suggests that tendencies towards misbehavior are intrinsic to human nature, rather than solely learned. This perspective is in alignment with Rom 5:12, that *all sinned in Adam* and that there is a transgenerational transmission of these tendencies, which is even supported by epigenetic mechanisms that influence how behaviors and propensities are passed down across generations. Tremblay emphasizes that intervention must begin early, even during pregnancy, to address these disruptive behaviors effectively. This illustrates the theological emphasis on the need for salvation to counteract the inherited propensity to sin. His conclusions about the importance of individualized approaches to addressing different subtypes of disruptive behavior problems add a scientific distinction to understanding how the original sin affects each person uniquely, requiring personalized approaches as well for correction and prevention. Tremblay's recent study supports a plausible epigenetic mechanism for transgenerational sin, suggesting that epigenetic processes play a role in the origins of disruptive behavior problems. This fascinating study highlights the impact of intensive perinatal interventions on various aspects of physical and mental health, providing good evidence that transgenerational epigenetic inheritance contributes to the transmission of mental, physical, and even spiritual disorders. By application to the transgenerational inheritance of sin, understanding that humanity was originally created in a state of righteousness, in the image of God, it was not inherently immune to sin. After the Fall, the changes to Adam's epigenome could have resulted in the inheritance of original sin through transgenerational epigenetic mechanisms, as described by Tremblay. In addition to Tremblay's work, as well as the results from other contemporary studies, this could explain the universal human tendency toward sinfulness.

EPIGENETIC FLEXIBILITY: THE POWER OF REVERSIBLE CHANGES

The key feature that sets epigenetics apart from genetics is that DNA methylation can be reversed; thus, the structure of a gene is generally permanent, whereas the function of a gene is variable—changeable. With that being stated, the changeable nature, or reversable nature of the function or expression of a gene, occurs through the action of the ten-eleven translocation (TET) enzymes, discussed previously. The TET enzymes mediate the removal of methyl groups from 5-methylcytosines, an action that effectually causes a silenced gene to become active—or expressed. This reversibility supports the idea that silenced genes can be activated following a significant event, such as trauma. Conversely, active genes could potentially become silenced, due to methylation of key regions on the gene, as in the example provided earlier with the glucocorticoid receptor gene being silenced after a prolonged duration of adversity. Extension of this biological process to theology, as described in the previous section, reversal of the genes that have been affected by original sin could reverse the course of depravity, and reduce our propensity to sin. This idea aligns with biblical teaching that, through a future dramatic event when we see Christ face to face, the effects of Adam's sin will ultimately be erased. As Paul wrote in his letter to the Corinthians, "now we see in a mirror dimly, but then face to face" (1 Cor 13:12). Even John the apostle adds that this will occur at Christ's second coming when "we shall be like him because we shall see him as he is" (1 John 3:2). These Scriptures allude to the fact that, without altering our human nature, we could be delivered from our sinful nature, potentially through the process of progressive sanctification. Such a transformation might be reflected in the reversibility of the epigenetic changes inherited from Adam.

The traumatic event of the Fall may have led to imprinted genes, with DNA methylation patterns or other epigenetic alterations persisting through generations. The Fall may have left its imprint on human DNA, influencing various epigenetic mechanisms, but it does not alter the structure of the human genome. This distinction is crucial in understanding epigenetic inheritance, where the effects of sin can be passed across generations, and the outcome of progressive sanctification revealed in eschatology, the study of humanity's ultimate destiny. In other words, as Christians anticipate Christ's return, judgment, and resurrection, epigenetic inheritance offers a way to understand how the spiritual consequences of sin might be altered across generations. The spiritual transformation of a Christian can then be explained by active as well as silenced genes leading to sanctification.

EPIGENETIC INHERITANCE: IMPLICATIONS FOR ESCHATOLOGY AND SANCTIFICATION

The intersection of eschatology and epigenetics initiates a spiritual and biological dialogue that offers deep insights into human transformation and sanctification. From an eschatological perspective, changes in epigenetic marks (specifically, methylation) between generations can indeed manifest in the mind, body, and soul, as described in 1 Thess 5:23: "Now may the God of peace Himself sanctify you completely; and may your whole spirit, soul, and body be preserved blameless at the coming of our Lord Jesus Christ." This passage supports the dual process of sanctification, wherein a follower of Christ is both *positionally* sanctified (Heb 10:10) upon salvation and *progressively* sanctified (Heb 10:14) through the ongoing work of the Holy Spirit. Positional sanctification refers to the believer's immediate sanctification upon accepting Christ, as mentioned in Heb 10:10: "By that will we have been sanctified through the offering of the body of Jesus Christ once for all." This signifies a definitive change in the believer's spiritual status. However, progressive sanctification, as described in Heb 10:14 ("For by one offering He has perfected forever those who are being sanctified") indicates an ongoing process of spiritual growth and maturity, which is facilitated by the Holy Spirit.

This dual process of sanctification reflects the concept of epigenetic changes. Just as a believer is initially sanctified and continues to grow in sanctification, epigenetic marks can be established and subsequently modified throughout one's life in response to various influences. This aligns with the concept of progressive sanctification, where the Holy Spirit's work in a believer's life can bring about positive transformation in the mind, body, and soul. The mind can be transformed through the renewal of one's thinking, as described in Rom 12:2: "Do not conform to the pattern of this world but be transformed by the renewing of your mind." This renewal process can be reflected in epigenetic changes that promote mental health and cognitive function. The body, as the temple of the Holy Spirit (1 Cor 6:19–20), also undergoes transformation. Healthy lifestyle choices, such as diet, exercise, and abstaining from harmful substances, can lead to positive epigenetic modifications. These changes not only improve physical health but also serve as evidence of the Holy Spirit's sanctifying work in a believer's life. The soul, encompassing the essence of one's being, is profoundly impacted by one's choices to live a life that pleases God, enabled by the power of the Holy Spirit. As believers grow in their relationship with Christ and are conformed to his image, this spiritual growth can be evidenced by changes at the epigenetic level, challenging the traditional dichotomy between the spiritual and the physical. It suggests that the work of the Holy Spirit in a believer's life has tangible effects that can be observed by positive lifestyle changes, evidenced by either gene expression or gene silencing. As will be discussed later, Christians who engage in regular prayer and the study of Scripture, and who participate in a faith community exhibit changes in brain structure and function, as well as improvements in mental and physical health, all of which could be rooted in epigenetic inheritance.

The Holy Spirit plays a central role in the process of sanctification as well by guiding and empowering believers to grow in their faith and be conformed to the image of Christ. Epigenetics provides a biological basis for this theological concept, that while genetic predispositions can influence behavior and health, they are not fixed. Environmental factors, such as lifestyle choices and spiritual practices, can lead to changes in gene expression, highlighting the potential for positive transformation. This aligns with the biblical view that believers are continually being transformed into the image of Christ. As stated in 2 Cor 3:18, "And we all, who with unveiled faces contemplate the Lord's glory, are being transformed into his image with ever-increasing glory, which comes from the Lord, who is the Spirit." This ongoing transformation involves not only spiritual growth but also changes in the mind, body, and soul. The Holy Spirit's work in a believer's life extends beyond spiritual transformation to include somatic, physiological, and behavioral changes and acknowledges that while the consequences of sin are real, they are not permanent. Through the ongoing work of the Holy Spirit, believers can undergo profound transformation in all aspects of their being, ultimately reflecting the image of Christ and the promise of complete sanctification at his return.

EPIGENETICS AS AN AFFORDANCE: ADVANCING SCIENCE BEYOND REDUCTIONISM

In recognizing the profound connections between biology, environment, and the Spirit, the reception of epigenetics within theological discourse has not been without its challenges. While reductionist approaches in science have greatly advanced our understanding of specific biological processes, they often fall short in capturing the complexity and deeper questions about the origins and relational aspects of living beings. These limitations risk reducing epigenetics to a purely mechanistic science, interpreting gene expression and behavior as deterministic processes while overlooking the value of human and spiritual experiences. Such a reductionistic perspective raises a crucial question: Are humans merely the sum of their biological responses, or does epigenetics reveal pathways through which creation, relationships, and Scripture shape not only our physical selves but also our spiritual and moral lives? In response to this question, it becomes very clear that there is a serious tension between reductionism and this novel perspective of epigenetics as a theological affordance.

As introduced in the preface and the cycle of redemption in this book, the concept of affordances, described by psychologist James Gibson (1979), can be explained as the possibilities for action that the environment provides to living beings.[48] Theologically reframed, affordances are invitations to engage with the richness of God's creation, the wisdom of Scripture, and the depth of human relationships as transformative resources. Epigenetics, understood as an affordance, moves beyond being seen as a series of genetic mechanisms, becoming instead a purposeful manifestation of God's grace, empowering human responsibility.

The implications of this perspective are vast. Consider, for instance, the evidence presented by Yehuda et al. that trauma can imprint itself epigenetically, creating a biological inheritance of suffering.[49] Does this capacity for imprinting also allow for the transmission of hope, healing, and redemption? If the scars of addiction can be imprinted into the epigenome, might the lives of faith, love, and service within a supportive community begin to overwrite those marks with patterns of restoration and resilience? This perspective not only challenges the deterministic view often tied to reductionist science, but also invites us to see how faith's redemptive power can transform both our bodies and souls.

By connecting epigenetics with theological affordances, God's grace is revealed in creation through the transformation of human beings—new creatures in Christ (2 Cor 5:17). As theologian N. T. Wright recognizes, humanity is "fearfully and wonderfully made" (Ps 139:14)—as embodied beings entrusted with the stewardship of creation and called to reflect the image of God.[50] This calling invites us to actively engage with creation, to live as agents of grace and restoration in a broken world. Augustine taught that sin's impact goes deeper than individual actions—it reaches the core of human nature. Epigenetics shows us how our environment shapes gene expression, and this insight can also help us understand the factors behind struggles like addiction. New research, such as that

48. Gibson, *Ecological Approach to Visual Perception*, ch. 8.
49. Yehuda, "Holocaust Exposure," 372–80.
50. Wright, *Simply Christian*, ch. 1.

by Dias and Ressler, reveals that positive experiences can reduce the inherited effects of trauma, offering an affordance of the theological promise that faith brings transformation and healing.[51] In integrating these insights, epigenetics, understood as a theological affordance, offers hope for those ensnared in addiction.

---

51. Dias and Ressler, "Parental Olfactory Experience Influences Behavior," 89–96.

5

Addiction

A Modern Reflection of Original Sin

ADDICTION CAN BE DEFINED as the compulsive use of a substance or activity, a craving, a behavior marked by repeated use despite destructive consequences.[1] Essentially, a pathological *wanting*. Addictive behavior occurs within the interchange of genetic and epigenetic factors. Understanding these interactions provides insights into the mechanisms underlying addiction and offers potential avenues for developing more effective treatments. From a neuroscientific perspective, addictive behavior is a persistent mental battle where neural mechanisms are associated with desire, compulsion, and lack of self-control or *behavioral inhibition*. A behavior based on the inability to extinguish unwanted behavioral responses, essentially a loss of behavioral inhibition, affects long-term memory.[2] Augustine alludes to the unwanted behavioral responses and social stigma seen in addictive behavior as he describes his aversion to vice in his *Confessions*,

---

1. Ruden, *Craving Brain*, 48.
2. Satel et al., "Addiction and the Brain-Disease Fallacy," 1–41.

> Nothing deserves to be despised more than vice; yet I gave in more and more to vice simply in order not to be despised.[3]

From these definitions, addiction (or vice) can be viewed as a habit, even a despised one, according to Augustine. Each of the definitions above share a commonality in their responsiveness to reason and therefore, their connection could be considered a voluntary action. The voluntary nature of addiction provides a strong defense for defining addiction more as a habit than a disease, particularly because the disease concept obscures the extent to which a person may be expected to take responsibility for their addictions. However, involuntary events, such as trauma, combat, or victimization, cannot be overlooked as root causes of addictive behavior. The idea that the innate potential for addiction or predisposition begins at the very start of life, with the union of sperm and egg to form a unique zygote, lies within this context. Combination of the DNA from both the sperm and egg establishes what has been referred to as a *biological terrain*.[4] A terrain is essentially the genes that have been inherited from both parents, evidenced by phenotypes, and it the case of addiction, addictive behavior. Just as waves continuously shape the sand on a beach, life experiences mold each person's individual genetic landscape, or in the context of epigenetics—*methylscape*. The methylscape is a terrain uniquely influenced by the environment, particularly in shaping a person's tendency toward addictive behavior. Although there are several genes that have been identified to play a role in addictive behavior, the expression of the genes is dependent on the genetic and epigenetic terrain that is unique for everyone. Individuals with heightened susceptibility to addiction—those whose addiction-related genes are strongly influenced by environmental cues—are more easily triggered.

The study of addiction has also centered around the social transmission of addictive behavior in family relationships that

---

3. Augustine, *Conf.* 2.3.
4. Ruden, *Craving Brain*, 26–29.

include identical twins, fraternal twins, adoptees, and siblings. The research findings have suggested that a significant contribution to an individual's risk of becoming addicted to substances such as nicotine, alcohol, or other drug depends on his or her genetic background. While a genetic component is evident, studies suggest that epigenetic changes in addiction-related genes offer critical insights into addressing drug addiction.[5] Uncovering the epigenetic mechanisms behind addictive behavior will also provide insights into the concept of epigenetic inheritance, which could also lead to the previous hypothesis regarding addiction's connection to original sin. To explore this hypothesis in greater detail, further details on the biological nature of addiction are needed, particularly how the expression of the genes might contribute to transgenerational epigenetic inheritance of addictive behavior, rooted in the concept of original sin.

WIRED FOR CRAVING: NEUROBIOLOGY AND THE CYCLE OF ADDICTION

Addictive behavior, particularly drug addiction, is one of the most pervasive and challenging issues of our time, impacting individuals, families, and communities on physical, emotional, and spiritual levels. Far from being a single choice or isolated incident, addiction gradually invades one's life through a series of interconnected stages. Each phase reflects a progression where underlying pain, unmet needs, and unresolved stressors create an ideal backdrop for harmful behaviors that ultimately bind the individual in a cycle of dependency. This cycle can be described in the account described in Jas 1:13–15:

> Let no one say when he is tempted, "I am tempted by God"; for God cannot be tempted by evil, nor does He Himself tempt anyone. But each one is tempted when he is drawn away by his own desires and enticed. Then, when desire has conceived, it gives birth to sin; and sin, when it is full-grown, brings forth death.

5. Lax et al., "DNA Methylation Signature of Addiction in T cells," 1–12.

Beginning with the *problems of life*, the trials and pressures that everyone encounters, can incite deep-seated fears and anxieties in certain individuals. When these feelings intensify, the desire to escape leads some to seek a *temporary sense of relief* through substances or behaviors that offer quick but short-lived comfort. However, the aftermath is often a overwhelming sense of *guilt and shame*, as the consequences of these actions begin to unravel relationships, health, and self-worth. This painful reality often causes the individual to *search for solutions* in a desperate attempt to manage or recover control.

Nevertheless, without addressing the root cause(s), such as the underlying emotional wounds and/or unmet spiritual needs, the individual is frequently drawn back into the familiar patterns of the Cycle of Addiction. In this cycle, the stages reflect James's illustration of how unchecked desire gives birth to sin and, if left unaddressed, ultimately leads to a kind of death or a separation from one's true self and from God. By understanding each of these stages, we gain insight into both the complexity of addiction and the necessity of confronting it at its roots, taking responsibility to engage spiritual practices and live a life of resilience, as presented in Figure 3.

Further exploration of the Cycle of Addiction reveals a complex pattern, stressing the challenges inherent in breaking free from its grip. Each phase feeds into the next, showing that recovery often requires interventions that address not only the symptoms, but the root causes of addictive behavior. The cycle is a repetitive and self-reinforcing process that begins with a trigger—an internal or external event that leads to cravings. These cravings drive the individual to engage in substance use or addictive behavior, seeking short-term relief. However, this temporary relief is soon followed by negative consequences, such as physical, emotional, or social harm. Feelings of guilt or shame often arise, which lead to withdrawal or distress, creating an internal struggle. This distress increases vulnerability to future triggers, restarting the cycle and making it difficult to break free without intervention.

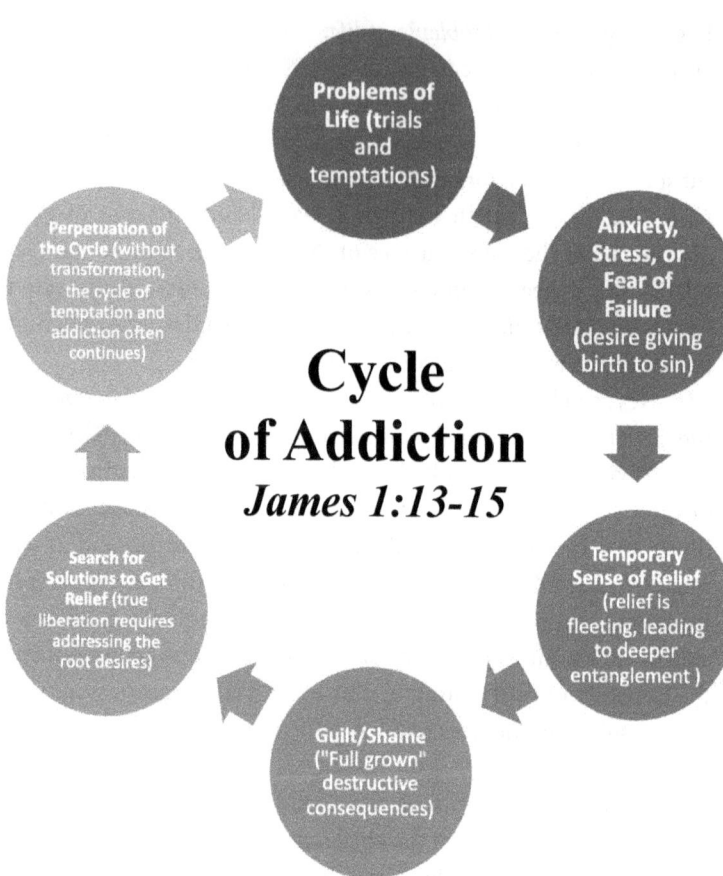

Figure 1. *The Cycle of Addiction.* The Cycle of Addiction is aligned with the progression of sin as described above in Jas 1:13–15, which illustrates how unaddressed struggles—whether past trauma, personal issues, or social pressures—create vulnerability to temptation. As James notes, God does not lead us to sin, but our responses shape our outcomes. Stress and anxiety amplify the desire to escape discomfort, giving way to temporary relief through addictive behaviors. This brief relief intensifies the entrapment, leading to guilt and shame as harmful consequences begin to unfold. Without confronting the underlying issues, attempts to escape the cycle often fall short, echoing James's caution that unchecked desire grows into sin, which, if unrestrained, leads to a form of death—a separation from one's true self and from God.

Key details of the cycle of addiction include developing a dependence on the drug, where the individual requires it to function normally. As tolerance builds, larger doses are needed to achieve the same effect. When the drug is unavailable in sufficient quantities, withdrawal symptoms occur, often leading to relapse. Addiction follows a cyclical pattern of compulsive behavior, frequently triggered by emotional factors like guilt or shame, which lead to mistrust and the need for immediate gratification, resulting in cravings that overpower the conscience.

With the emotional brain activated, cognition is then engaged with a habit or ritual motion toward acting out the preconceived action. Although the "acting out" creates excitement, the satiation is limited by feelings of guilt, shame, and regret. The neural strengthening of the event that has occurred is facilitated by a temporary reassurance and reminder based on an emotional trigger. Hence the painful cycle of addiction continues, which alters physiological systems and contributes to the maintenance of the addictive state and influences withdrawal and relapse.[6] The perpetuation of the cycle is dependent on the substance, such as opiates, which help to remove not only physical pain but also emotional pain. The anterior cingulate region of the brain, which is responsible for relationality and attachment or bonding with others, produces the affective aspects of pain. It is a region that is activated in the addiction cycle following social rejection or social isolation. Interestingly, there is a dramatic increase in the concentration of neurotransmitters in the anterior cingulate during moments of human bonding and also in the emotional response to pain.[7]

It is noteworthy also that the emotional response to pain resides in a different region of the brain than that of physical pain. This combination of attachment and affective pain may explain the benefits of being held while in physical pain. However, studies in juvenile animals that have been subjected to intense pain have demonstrated that when the anterior cingulate is saturated with opiates, the animals become despondent and stop seeking

6. Nielsen et al., "Epigenetics of Drug Abuse," 1149–60.
7. Waters, *Addiction and Pastoral Care*, ch. 3.

comfort from their mothers.[8] These results suggest that the opiate has facilitated their ability to remove emotional as well as physical pain, which may provide a rationale for the continued use of opiates by traumatized individuals who seek to medicate pain from childhood maltreatment or trauma.

Opioids, the most prominent of all addictive substances, remove not only physical pain and psychological distress, but also the emotional desire to relate to others. Opioid addiction isolates us in profound ways, mimicking the feeling of social connectedness with humans and habituating the user to seek social fulfilment from drugs.[9] Those ensnared by opioid abuse cease to find God and humans as sources of consolation, which creates a serious spiritual problem as it has the capability of causing harm to the soul. One of the greatest assets in the Christian life for overcoming this is the transformation that comes from the renewing of the mind by the power of the Holy Spirit (Rom 12:1–2). This transformation is further strengthened when recovering individuals are in a community of fellow believers who are also striving to live sanctified lives.

The profound influence of the environment on individuals recovering from addiction was first highlighted in Bruce Alexander's landmark 1970s study, well known as *Rat Park*.[10] Results from these studies concluded that the environment is critical in influencing human drug use, and that a complex interaction of individual risk (genetic and environmental), integrated with a larger social system, are essential constructs of addictive behavior. The individual biological factors and social interactions that were revealed in the *Rat Park* findings have brought clarity to the idea that individual biological factors and social interactions may be more of a genetic influence than what traditional nature–nurture dichotomy studies would indicate. In the context of drug addiction, however, the interactions between genetic and environmental

---

8. Inagaki, "Opioids and Social Connection," 85–86.
9. Inagaki, "Opioids and Social Connection," 87–90.
10. Gage et al., "Rat Park," 917–22.

factors point toward an important role for epigenetic mechanisms in the acute response to drugs and the development of addiction.[11]

INHERITANCE OF ADDICTION: GENETIC AND EPIGENETIC

Addiction can be considered an *idiopathic* condition, mainly because it is difficult to understand why certain individuals are more susceptible to developing addictive behavior than others. Even with evidence that stressful life events are risk factors for the development of addiction, not all individuals who have experienced highly stressful events develop addictive behavior. Individuals with a greater susceptibility to addiction are influenced not only by stress but also by factors such as family dynamics, moral development, and socioeconomic context. Although these extrinsic factors can be detrimental to addictive behavior, a genetic pre-disposition to addiction also plays a crucial role. It has been determined that 40–60 percent of the population is genetically prone to developing an addiction.[12] With such a high percentage of susceptibility of developing an addiction, specific genes have been identified that influence individuals towards certain types of substances.[13] It is important to note, in this context, that environment plays a major role in the expression of a variety of genes related to addiction.[14] Early life adversity, stressful life events and lower levels of education all seem to have a major effect on the genes involved in alcohol metabolism and neurotransmitters, dopamine and serotonin.[11]

In addition to genetic alterations that play a role in the development of addiction, epigenetic alterations to these genes can lead to addiction to psychostimulants.[15] Furthermore, repeated stressful life experiences, particularly stress-related abnormal

11. Cunliffe, "Epigenetic Impacts of Social Stress," 1653–69.
12. Koob, "Neurobiology of Addiction," 760–73.
13. Jacqueline, "Genetics of Addiction," 684–87.
14. Trifu, "Aggressive Behavior in Psychiatric Patients," 3483–87.
15. Cadet, "Transcriptional and Epigenetic Substrates," 696–717.

maternal behavior, can result in significant epigenetic changes in the glucocorticoid receptor gene, NR3C1.[16] This gene and the brain-derived neurotropic factor gene BDNF are two prominent genes that are at the root of addictive behavior. Epigenetic alterations to the NR3C1 and BDNF genes due to stressful life experiences have been studied extensively in their role in addiction and trauma.

Glucocorticoid Receptor Gene, NR3C1

The glucocorticoid receptor gene, NR3C1, is a crucial gene that encodes the glucocorticoid receptor (GR) protein responsible for binding to glucocorticoid hormones, such as cortisol, in the body. This receptor plays a central role in the body's response to stress, regulation of the immune system, and various other physiological processes. Stress is a known trigger for addictive behaviors. The glucocorticoid receptor is a transcription factor that regulates gene expression in response to glucocorticoid hormones. When glucocorticoid hormones, such as cortisol, are released in response to stress or inflammation, they bind to the glucocorticoid receptor, which then translocates into the cell nucleus and can modify the expression of target genes.[17] Epigenetic modification of the methylscape within the promoter region of the NR3C1 gene is one of the most common alterations. The methylation pattern can either increase or decrease gene expression because of stress in the context of addiction. When there is increased DNA methylation at the NR3C1 gene promoter region, there is decreased expression of the glucocorticoid receptor. This can disrupt the stress-response system, making an individual more vulnerable to stress and potentially increasing the risk of addiction. Even when DNA methylation is reduced at the NR3C1 promoter region, there is a higher level of the glucocorticoid receptor expressed, which can lead to an overactive stress response that could contribute to addiction

16. Walker, "Maternal Touch and Feed as Critical Regulators of Behavioral and Stress Responses," 638–50.

17. Oakley et al., "Biology of the Glucocorticoid Receptor," 1033–44.

by influencing an individual's susceptibility to stress-induced cravings and relapses.

It is evident from these conditions that a balance of DNA methylation in the promoter region of the glucocorticoid gene is necessary for maintaining the stress related to environment. Epigenetic modifications to the *NR3C1* gene have also been studied in the DNA obtained from umbilical cord cells in newborns exposed to maternal depression.[18] The results demonstrated that prenatal exposure to increased third-trimester maternal depressed/anxious mood was associated with increased methylation at the promoter region of the *NR3C1* gene. The increased methylation at the promoter region was also associated with increased cortisol stress responses at three months, indicating that the increased cortisol was due to maternal stress as evidenced in the progeny. In a particular study of the postmortem brain tissue obtained from suicide victims who had a history of child abuse, the promoter region of the *NR3C1* gene was also highly methylated, indicating very low levels of the glucocorticoid receptor.[19] The extent to which the promoter region of the *NR3C1* gene can be altered by a highly stressful environment was exemplified by survivors from the Rwanda genocide in 1994. A genome-wide evaluation of the genes affected by the extremely traumatic event demonstrated an increase in the methylation at the *NR3C1* promoter region in DNA obtained from leukocytes of genocide survivors.[20] From these studies it is evident that stressful events—including maternal deprivation and trauma associated with genocide—can cause long-lasting epigenetic changes that are measurable at all stages of life in humans.

18. Oberlander et al., "Prenatal Exposure to Maternal Depression," 97–106.

19. McGowan et al., "Epigenetic Regulation of the Glucocorticoid Receptor in Human Brain," 342–48.

20. Vukojevic et al., "Epigenetic Modification of the Glucocorticoid Receptor Gene," 10274–84.

Brain-Derived Neurotrophic Factor Gene, BDNF

Another major gene associated with addiction, particularly due to stressful events that can affect epigenetic modifications of the methylscape, is the hippocampal *brain-derived neurotrophic factor* (*BDNF*) gene. Levine et al. have suggested that addictive behavior can influence the epigenetic regulation of the *BDNF* gene.[21] As an essential neurotrophin, BDNF plays a significant role in neuronal survival, development, and synaptic plasticity. Epigenetic modifications associated with addictive behavior occur at specific regions within the *BDNF* gene promoter regions, such as chromosomal exons I and IV. These regions have increased DNA methylation and have been associated with substance abuse, alcohol dependence, and nicotine addiction.[22] Higher methylation levels in these regions are generally linked to decreased *BDNF* expression. Environmental factors, such as chronic drug exposure or stress, can induce epigenetic modifications to the *BDNF* gene, leading to changes in *BDNF* expression levels. For example, exposure to drugs of abuse or chronic stress has been associated with increased DNA methylation as well as other epigenetic changes at specific *BDNF* gene regions.[23]

It is important to note that the exact epigenetic changes to the *BDNF* gene resulting from addictive behavior may vary depending on factors such as the type of addiction, duration and intensity of substance use, and individual variations. Of relevance to this gene, is a study conducted by Varghese et al. that demonstrates how BDNF plays a vital role in brain health, as it supports synaptic plasticity, neuronal growth, and the overall survival of brain cells. In the context of mental health, individuals suffering from conditions like depression may find improvement through faith-based

---

21. Levine et al., "Epigenetic Mechanisms of Addiction," 424–35.

22. Klengel et al., "Role of DNA Methylation in Stress-Related Psychiatric Disorders," 115–32.

23. McGowan et al., "Epigenetic Regulation of the Glucocorticoid Receptor," 342–48.

activities, which have been shown to elevate serum BDNF levels.[24] These activities influence key neural systems involved in addiction, which include the hypothalamic-pituitary-adrenal (HPA) axis and the reward circuitry, particularly the ventral tegmental area and nucleus accumbens.[25] Animal models have also shown that chronic stress can reduce BDNF levels in regions like the prefrontal cortex (PFC) and hippocampus, which are key areas linked to mood regulation seen in postmortem studies of depressed suicide victims.[26] Interestingly, it has been shown that increased levels of BDNF are associated with "high intrinsic religiosity" among depressed patients.[27] This finding would suggest that elevation of BDNF levels could be a potential method for protecting spiritual experiences, such as prayer, in depressive disorders.

Kappa Opioid Receptor Gene, KOR or OPRK1

The kappa opioid receptor gene (*KOR* gene), often referred to as *OPRK*1, encodes the kappa opioid receptor, which is part of the body's opioid system.[28] This receptor interacts with various naturally occurring chemicals in the brain called *dynorphins*, which are endogenous opioid peptides. Dynorphins bind to kappa opioid receptors to modulate pain, stress responses, and emotional regulation. The kappa opioid receptor has several key roles in regulating mood, stress, and addiction-related behaviors. *KOR* activation generally produces analgesic (pain-relieving) effects. It can also lead to dysphoria (feelings of unease or dissatisfaction) rather than

---

24. Varghese et al., "Role of Spirituality in the Management of Major Depression," 1–2.

25. Miao, "Relationship Between Stress, Mental Disorders," 1–20.

26. Pandey et al., "Brain-Derived Neurotropic Factor and Tyrosine Kinase B Receptor," 1047–61.

27. Mosqueiro et al., "Increased Levels of Brain-Derived Neurotropic Factor," 671.

28. Cortes et al., "Does Gender Leave an Epigenetic Imprint on the Brain?" 3.

the euphoria typically associated with other opioid receptors like the mu-opioid receptor.

The KOR signaling pathways play a role in mediating negative emotional states, stress responses, and resilience to traumatic events. Particularly relevant to the gene-environment paradigm in addiction, KOR activation is involved in reducing reward-seeking behaviors, as its stimulation can diminish dopamine release in the brain, leading to less pleasure and craving, particularly in relation to substances like alcohol and drugs. This receptor is of intense research as it has shown that early childhood abuse and trauma can significantly affect the regulation of the KOR system. Chronic stress and trauma activate the body's dynorphin system, which, when bound to KOR, amplifies negative emotional responses. This can lead to heightened stress sensitivity and dysphoric states in individuals who experienced abuse early in life. Overactivity of the KOR due to early trauma can contribute to emotional dysregulation and the development of anxiety, depression, and post-traumatic stress disorder (PTSD). Individuals with heightened KOR activity may experience increased sensitivity to stress and reduced capacity for emotional resilience.

Many of these alterations due to early abuse may also increase vulnerability to substance abuse later in life. Given that KOR activation is linked to negative mood states, individuals may seek external substances (e.g., alcohol, drugs) to counteract the dysphoric feelings associated with overactive KOR signaling. In agreement with studies conducted by McGowan on the BDNF gene, early life stress can lead to epigenetic modifications of the KOR gene, altering how the gene is expressed. Such changes might affect how the KOR gene functions in response to stress and trauma throughout life, as epigenetic modifications in the gene can lead to maladaptive stress responses, increasing the risk of mental health disorders and addiction.

Additional Genetic and Epigenetic Drivers of Addiction

Addictive behavior is shaped by the interaction of many other genes as well, each playing a critical role in susceptibility to addiction. Among the more influential ones are the dopamine receptor genes (D1–D5), which regulate dopamine neurotransmission, a process integral to mood, reward, motivation, pleasure, and various cognitive functions. These receptors help control movement, reinforce rewarding behaviors, and inhibit certain neuronal activities linked to addiction. Additionally, genes like *SLC6A3* (*DAT1*), which regulates dopamine transport, and *SLC6A4* (*5-HTTLPR*), which influences serotonin transport, also affect dopamine pathways and addiction risk. Specific to alcohol addiction, the *GABRA2* gene, which codes for a subunit of the GABA-A receptor involved in inhibitory neurotransmission, plays a key role, while the *NYP* gene, associated with stress and anxiety regulation, is linked to alcohol dependence and other addictions. Notably, these genes—and many others—are not only pivotal in neurotransmitter systems but are also regulated by epigenetic mechanisms such as stress or substance exposure, which underscores the multifactorial nature of addiction.

ADDICTION UNRAVELED: GENETIC PREDISPOSITION MEETS ENVIRONMENTAL TRIGGERS

While genetics play a significant role, environmental, social, and psychological factors are also crucial in its development and progression of addiction. Understanding gene–environment interactions, such as how stress or drug exposure can lead to epigenetic changes, individuals with certain genetic variants may be more susceptible to epigenetic changes induced by drug use. In addition to the gene–environment relations are the gene–gene interactions in which genetic variants can influence each other's expression through epigenetic mechanisms. When a variant in a dopamine receptor gene, for example, alters the methylation status of a transporter gene, this could modify how the dopamine receptor gene is expressed. Epigenetic changes in the *DRD*2 gene, such as increased methylation, have been linked to reduced receptor expression in individuals with addiction. This leads to altered dopamine signaling and increased susceptibility to addictive behavior. Stress can also affect the function of essential behavioral genes, such as the *CRF* (Corticotropin-Releasing Factor) gene. Stress can induce epigenetic changes in the *CRF* gene, increasing its expression and contributing to stress-induced relapse in addiction. Additionally, variations and epigenetic modifications in the *GABRA*2 gene have been linked to alcohol dependence.

By the combination of behavioral therapies with epigenetic treatments, therapeutic interventions are currently in progress to enhance the efficacy of addiction treatment by addressing both genetic predispositions and environmental influences. Epigenetic modifications can be immediate or accumulate slowly and may be passed on to successive generations via germ (reproductive) cells or remain with an individual via somatic (body) cells. These epigenetic alterations may be due to inheritance through genomic imprinting, prior life events, chronic drug use or pharmacotherapies used for the treatment of addictions.[29] The causal nature of addictive behavior cannot be adequately appraised until it is understood from other perspectives than neurobiology. Neurobiological accounts alone are limited, particularly in their attempts to explain strategies for overcoming addictions without medical treatment. Addicted individuals in recovery programs who seem to experience frequent relapses rarely change their social, occupational, psychological, and economical environments, as evidenced by their lack of personal resilience. By the mental strengthening that occurs in resilience training, as discussed in the final chapter, epigenetic modifications may occur that reinforce stability of personal freedom and will, particularly under the stresses that may lead to relapse.

Environmental exposures or even choices people make can remodel the methylscape that may affect multiple genes. Although each cell type in the human body effectively contains the same

---

29. Nielsen et al., "Epigenetics of Drug Abuse," 1149–60.

genetic information, epigenetic regulatory systems enable the development of different cell types (e.g., skin, liver, or nerve cells) in response to the environment. These epigenetic changes can affect health and even the expression of the traits passed to children. For example, when a person uses cocaine, epigenetic changes to DNA, such as methylation, increase the production of proteins that are associated with addictive behavior. Increased levels of these altered proteins correspond with drug-seeking behaviors that have been evaluated in animals. Epigenetic studies conducted with identical twins, both born with the same DNA sequence, have demonstrated that, after birth, as the twins mature and are exposed to differences in their environments, they begin to make choices of their own. As a result, over their lifetimes, the differences in their methylscape are evidenced by their behavior, specifically their risk of addiction, and even their response to treatment. Some of these changes can be transgenerational, as has been described for trauma. Studies in transgenerational trauma may provide an explanation for, and potential solution to, transgenerational addiction.

As a current member of the pharmaceutical community, I am certainly not in opposition to the use of drugs to ameliorate the addictive nature of illicit substances. Continued neurobiological research on addiction is essential for our understanding of the direct effects of drugs on the brain. The development of drugs that block or suppress epigenetic, transcriptional, and biochemical changes associated with sensitivity to stress may have a significant impact on the lives of individuals suffering from addiction. Although complex and common diseases are finally within reach of genetic medicine, the new frontier in genetic medicine is epigenetic therapies that are destined to transform the treatment of of many diseases, including cancer. Only a fraction of the genetics associated with disease involves changes in protein-coding genes. Many human diseases result from alterations in gene expression, driven by epigenetic regulators that can control a cell's state and fate. Epigenetic therapy can change cell states at will, without causing damage to the DNA sequence. Subtle changes in epigenetic networks determine how cells are specified and shifted between healthy, diseased, exhausted, and rejuvenated states. By reshaping these networks, genes can be activated or repressed to reassign cellular states and alter the course of a disease. The epigenome controls levels of gene expression, as mechanisms either increase or restrict the ability of *transcription factors* to read DNA. By altering DNA methylation or demethylation, the binding sites of transcription factors can be modified, meaning that the "on" and "off" switches of a gene can be moderated or tuned so that the protein can be expressed correctly. Other modifications that alter the shape of DNA (such as histone modifications) can act synergistically with DNA modifications to regulate expression.

Epigenetic therapies, specifically epigenetic editing, aim to modulate gene activity by precisely modulating gene expression transiently or persistently without impacting genomic integrity, presenting a therapeutic path to a wide range of diseases from rare genetic disorders to complex and common illnesses. An epigenetic therapy includes two components: 1) highly specific DNA binding domain (DBD)—e.g., dCas9 with guide RNA or zinc finger protein—that can target one or more selected genes; 2) effector(s) that can activate or repress gene expression. Exquisite specificity is conferred by the DNA binding mechanism and the specific interaction between the DBD binding site and its effector(s). In contrast to gene-editing therapies that require the breakage of DNA, epigenetic therapies modulate gene expression without posing translocation or chromosomal rearrangement risks. In addition, they can be extensively multiplexed to achieve additive and/or synergistic effects. The delivery mode of epigenetic therapies is generally selected after considering the indication and target tissue, and can include viral or non-viral approaches. A vast majority of diseases are driven by epigenomic regulators that control gene expression and cell state. By focusing on epigenomic control elements, which make up most of the human genome, epigenetic therapies can pioneer a therapeutic path for a host of common and rare diseases.

6

Original Sin and Addictive Propensity
Augustine's Theological Perspective

AUGUSTINE'S BASIS FOR ORIGINAL SIN is that humanity is in solidarity with Adam and that we have inherited a propensity towards sin since humans are *born in sin*; as the psalmist laments, *in sin my mother conceived me* (Ps 51:5). Paul affirms the universality of sin, noting that we are "by nature children of wrath" (Eph 2:3) and that "all sinned in Adam" (Rom 5:12), a truth later emphasized by Augustine. The state of original sin leaves us in the wretched condition of being unable to refrain from sinning. We still can choose what we desire, but our desires remain confined by our sinfulness. He asserted that the freedom that remains in the will always leads to sin. Therefore, in the flesh we are free only to sin, which is freedom without liberty, leaving us all in serious moral bondage.[30] As Paul lamented, sin cannot be defeated and won because there is no power in himself, other than himself, to stop sinning (Rom 7:19). Paul found himself caught in the desperate powerlessness of trying to battle sin in the power of *self*. In his sermon "How Sin Makes Us Addicts," Tim Keller defines sin as the act of replacing God with something or someone else, resulting in an addiction of the spirit.

---

30. Augustine, *Conf.* 7.3.

> There is an attraction at the spiritual level every bit as powerful as sexual attraction at the physical level: You cannot produce your own meaning in life, your own worth, your own security. Spiritually speaking, if it's not God who is the source of your meaning, then you're in bed with something else.[31]

This contemporary understanding of sin sheds light on Augustine's view of sin as the bondage of the human will. It demonstrates how sin as an act, as a habit, and as a pre-disposition, mirrors the testimonies of addicts—that their behavior is at one and the same time voluntary, and yet beyond the immediate control of a supposedly autonomous will.[32] From Augustine's *Confessions* emerges a vivid description of his struggle with sin, which is an extraordinary depiction of addictive behavior:

> I was bound by the iron chain of my own will. The enemy held fast my will, and had made of it a chain, and had bound me tight with it. For out of the perverse will came lust, and the service of lust ended in habit, and habit, not resisted, became necessity. By these links, as it were, forged together—which is why I called it a chain—a hard bondage held me in slavery. But that new will which had begun to spring up in me freely to worship you and to enjoy you, my god, the only certain joy, was not able yet to overcome my former willfulness, made strong by long indulgence. Thus, my two wills—the old and the new, the carnal and the spiritual—were in conflict within me, and by their discord they tore my soul apart.[33]

Further to his dilemma with sin, and a mirror to one who is at war with an addiction, Augustine finds himself to have a willingness to be what he was unwilling to be:

---

31. Keller, "How Sin Makes Us Addicts."

32. Tremblay, "Developmental Origins of Disruptive Behavior Problems," 341–67.

33. Augustine, *Conf.* 8.5.10.

> I truly lusted both ways, yet more in that which I approved in myself than in that which I disapproved in myself. For in the later it was not now really, I that was involved, because here I was rather an unwilling sufferer than a willing actor. And yet it was through me that habit had become an armed enemy against me, because I had willingly come to be what I unwillingly found myself to be . . . I hesitated to give up the world and serve you because my perception of the truth was uncertain. For now, it was certain. But, still bound to the earth, I refused to be your soldier, and was as much afraid of being freed from all entanglements as we ought to fear to be entangled."[34]

Even at the pinnacle of despair, Augustine contrasts love with lust.

> Your love satisfied and vanquished me, my lust pleased and fettered me.[35]

This conflict provides insight into addiction as a *conflict of desires*, described in terms of the *divided will* or *self*. Although a conflict of desires and a divided will can certainly apply to every human, the core idea is that addiction can often become a spiritual situation that affects not only the body, but the body, soul, and spirit.[36] In his commentary on Augustine's *City of God*, William Babcock found that Augustine argued that sin must involve a free exercise of will, otherwise it will not count as the agent's own act for which the agent is morally responsible.[37] In another study of *City of God*, James Wetzel claimed that Augustine associated involuntary sin with habit.[38] These statements could be misleading, however, depending on how one understands the term *habit*. While a habit can refer to one's natural constitution, Babcock and Wetzel refer to a habit as *tendencies built up over time*. Augustine acknowledges that tendencies developed in one's life can have the compulsive

---

34. Augustine, *Conf.* 8.5.11.
35. Augustine, *Conf.* 8.5.12.
36. Cook, *Alcohol, Addiction and Christian Ethics*, ch. 6.
37. Babcock, "Ethics of St. Augustine," 116
38. Wetzel, *Augustine's City of God*, ch. 2.

force that Babcock and Wetzel emphasize. Yet in his later writings Augustine emphasizes that sin's power is not mainly due to choices made along an individual's path in life, but rather is due to carnal concupiscence, which is a constitutional fault. It was Augustine's opponent Julian who argued that Paul referred to the *body of death* as habitually sinful, meaning that any struggle with sin is a struggle with bad habits, not with a sinful nature.[39] Augustine replied that the mortal body of death that Paul referred to was due to the *law of sin* in the corruptible body, not habit.[40] He does not deny that habits can lead us to sin, even when unwilling, but he insists that evil does not become ingrained through the force of personal choices that harden into habits.[41] As soon as persons begin to have the use of reason, they already experience the concupiscence of the flesh, which had been asleep because of their young age, as if waking up and fighting back.[42] Augustine gives a theological account of human sinful willing that seems to match the dynamics plainly at work in addiction.

A key similarity of original sin to addiction is that we can choose what we desire, but our desires remain chained by our evil impulses. As one can conclude, the dilemma of bondage to sin is nearly identical to the testimonies of those who are tormented with addictions. Augustine argued that the freedom that remains in the will always leads to the bondage of sin, a progression that is very familiar to those who struggle with addictions. However, this condition indicates that all of us are totally dependent upon grace for our conversion.

Given the description of Augustine's view of original sin and its potential parallels with addictive behavior, we can explore how epigenetic changes in genes related to addictive behavior may be viewed within this paradigm. While Augustine did not have knowledge of epigenetics, we can consider how his concepts of inherent human weakness and the struggle for redemption align

---

39. Augustine, *C.Jul.imp.* 1.67.
40. Augustine, *C.Jul.imp.* 1.69.
41. Augustine, *C.Jul.imp.* 1.72.
42. Augustine, *C Jul.imp* 3.178.

with our understanding of epigenetic mechanisms and their influence on addictive behavior. Epigenetic changes can occur in genes associated with reward pathways, stress responses, impulse control, and other processes implicated in addiction.[43] With these epigenetic alterations in mind, could it be that humans originally had unaltered DNA (Gen 1:31, Deut 32:4), and that the behavioral genes were evidenced by a methylscape that demonstrated a stress-less life in complete harmony with God? Yet, following the disobedient act in the garden, could it be that the original sin affected the methylscape in such a way that this traumatic incident resulted in imprinted epigenetic changes in the first human's genome?

In Augustine's view, original sin introduces a fundamental flaw in human nature, resulting in a state of spiritual and moral bondage. This perspective aligns with the concept of addiction as a manifestation of disordered desires and the loss of autonomy. Epigenetic changes in genes related to addictive behavior can be seen as reflecting the consequences of original sin at a biological level, further contributing to the inherent human weakness and susceptibility to addiction. As discussed previously, studies have suggested that epigenetic modifications in genes encoding dopamine receptors, such as *DRD2* and *DRD4*, may influence reward processing and contribute to addictive behaviors.

Augustine's understanding of original sin emphasizes the brokenness and disordered desires in human nature, which can be paralleled with the dysregulation of reward pathways observed in addiction. Epigenetic changes in these genes may further disrupt the normal functioning of reward systems, exacerbating a vulnerability to addictive behaviors. Additionally, genes involved in stress responses, such as the corticotropin-releasing hormone receptor 1 (*CRHR1*) gene, have also been implicated in addiction.[44] Epigenetic modifications in *CRHR1* have been associated with altered stress reactivity and an increased risk for substance abuse.

---

43. Nestler, "Epigenetic Mechanisms of Drug Addiction," 259–68.

44. Philibert et al., "Serotonin Transporter mRNA Levels are Associated with the Methylation," 844–49.

Augustine's view of original sin recognizes the impact of concupiscence, disordered desires, and passions on human behavior. Epigenetic changes in stress-related genes may reflect the disrupted stress response systems that contribute to addictive behaviors, further aligning with Augustine's perspective on the effects of original sin.[45] Moreover, Augustine's emphasis on the need for grace and divine intervention in overcoming the effects of original sin can be related to the role of epigenetic mechanisms in addiction recovery. Mercifully, epigenetic modifications are reversible and can be influenced by environmental factors, including interventions and therapeutic approaches. Just as Augustine asserts the importance of divine grace in redemption, the concept of epigenetic plasticity suggests that interventions—such as behavioral therapy, support systems and environmental changes—can modify epigenetic marks and potentially restore normal gene expression patterns associated with healthy behaviors.[46]

While Augustine's view of original sin does not directly account for the specific mechanisms of epigenetic changes, it provides a theological perspective reflected in the consequences of addictive behavior and the potential influence of epigenetics on addiction-related genes. By considering inherent human weakness and spiritual struggle in Augustine's view, we can draw connections to the disrupted biological processes observed in addiction and the potential role of epigenetic changes in contributing to addictive behavior. Further, it may be possible to understand the transgenerational effects of addiction based on the theological framework of original sin.

TRANSGENERATIONAL EPIGENETICS: THE LEGACY OF SIN AND ADDICTION

From the book of Exodus, the sins of the fathers are visited upon the children to the third and fourth generations (Exod 20:5-6).

45. Nestler, "Epigenetic Mechanisms in Psychiatry," e13-e14.

46. Volkow et al., "Relationship Between Subjective Effects of Cocaine and Dopamine," 827-30.

Long dismissed as theological allegory, this notion now finds an intriguing analogies in the science of molecular biology. The concept of transgenerational epigenetic inheritance—where environmental experiences can leave heritable marks on our DNA—offers a revolutionary perspective on the continuity of life's challenges, behaviors, and even traumas. For decades, scientists believed that each new generation began with a genetic *tabula rasa*, a blank slate, of epigenetic modifications that have been wiped clean during gametogenesis and early embryonic development. However, recent research challenges this dogma, in that DNA methylation and histone modifications are not always erased, as we have described previously. In some cases, they persist by epigenetic imprinting, serving as a molecular ledger of life's adversities and adaptations, inherited images of the past that become a biological narrative of future generations.[47] In the studies described previously in rodents, compelling evidence has been demonstrated that maternal stress, dietary deprivation, or exposure to toxins, induce specific epigenetic changes in the germline. These changes influence the stress responses, metabolism, and even the behavior of offspring.[48] For instance, male mice subjected to chronic stress exhibit altered DNA methylation patterns in their sperm, which correlate with heightened anxiety-like behaviors in their progeny. Similarly, experiments with dietary restrictions reveal that nutritional deficits can reshape methylation patterns in ways that predispose offspring to metabolic disorders. These findings suggest that life's hardships are encoded not just in memory or culture but in the molecular scaffolding of life itself.

Humans, too, bear the imprints of ancestral experiences, evidenced by the Dutch Hunger Winter of 1944–1945, a period of severe famine during World War II, which serves as a natural experiment in epigenetic inheritance. Individuals conceived during this famine exhibit distinct epigenetic markers related to metabolic

---

47. Yehuda et al., "Public Reception of Putative Epigenetic Mechanisms," 1–7.

48. Vassoler, "Impact of Exposure to Addictive Drugs on Future Generations," 269–75.

and cardiovascular diseases—markers that persist decades later and influence subsequent generations. Such discoveries reveal that the body not only records but also transmits responses to environmental stressors across generational lines. What mechanisms underlie this mysterious inheritance? Enzymes like TET1 and JMJD3 are central players. TET1, a DNA demethylase, facilitates the removal of methyl groups from DNA, shaping gene expression in response to environmental cues. JMJD3, a histone demethylase, unlocks genes silenced by repressive histone marks, enabling cellular reprogramming during critical periods of development. Together, these enzymes create a flexible epigenetic landscape, allowing the genome to adapt while retaining the imagery of past experiences. Remarkably, their imprint in germline cells suggests a route through which epigenetic *memories* evade the generational reset.

The implications are profound, in that our genes carry not only the instructions for building life but also the stories of lives lived. This perspective also reshapes our understanding of morality, health, and responsibility. If behaviors and conditions such as addiction or trauma are influenced by epigenetic inheritance, then the burdens we bear—and the actions we take—may be shaping the genetic and emotional legacies we leave behind. In the context of *Genes of Eden*, transgenerational epigenetic inheritance of original sin proposes that the past is never truly past, that human dispositions and struggles are deeply rooted in inherited conditions. Where Augustine saw the spiritual imprint of sin, epigenetics reveals molecular imprints encoded in our *epigenome*. Together, they underscore a profound truth: our lives are woven into the fabric of our epigenome, which spans generations, carrying the imprints of our ancestors while leaving traces for those yet to come.

The first suggestion of transgenerational epigenetic inheritance of Adam's sin can be found in Irenaeus's *Against Heresies*, in which his *doctrine of recapitulation* explained the notion of Christ's role as the Last Adam, who came to undo the curse brought upon humanity and the world by the first Adam's sin.[49] Shortly thereafter, two theories were put forward to explain the *hereditary taint* and the mode of the propagation of sin, and also an account for the implicit participation of the human race in Adam's sin.[50] Tertullian's Stoic-influenced traducianism and Origen's tradition of infant baptism, which viewed birth as inherently impure under the Law, inspired Irenaeus to formulate the idea of sin as hereditary. Apparently, he arrived at the truth of racial solidarity, as expressed by Paul, and then proceeded with the concept of mankind's potential (seminal) existence in their first father, just as the writer of the Epistle to the Hebrews regarded Levi as existing *in* Abraham (Heb 7:5). From Augustine's theory of the transmission of original sin, the primal sin of Adam is deterministic, meaning that all his progeny will be imprinted with the original sin.

Augustine also includes a second element of social transmission, which emphasizes that an individual's act of sin will have effects on their children.[51] These ideas support a materialistic basis for social transmission of behaviors within our epigenome that could be explained by transgenerational epigenetic inheritance.[52] Humanity's fall into sin, Augustine argued, was initiated through freedom—a freedom that, as a result of divine punishment, became corrupted and transformed into a bondage of necessity.[53] This corruption of human nature gave rise to a propensity to sin, passed down through generations not as genetic material but as an inherited inclination rooted in the human condition.

Pelagius, by contrast, denied any inherited propensity to sin, maintaining that Adam's sin impacted only Adam and that all subsequent human sin arises solely from individual free will. However, such a view fails to account for the universal tendency toward sin that humanity displays. While created in the image of God, in a state of original righteousness, humans were endowed

---

49. Irenaeus, "Against Heresies," 442–48.
50. Couenhoven, "St. Augustine's Doctrine of Original Sin," 359–96.
51. Augustine, *C.Jul.imp.* 6.21, 23.
52. Szyf et al., "Social Environment and the Epigenome," 46–60.
53. Augustine, "Perfection of Human Righteousness," 4.9.

with the ability to sin—and this capacity became a reality with Adam's fall. Original sin leaves humanity in a state of moral and spiritual incapacity, unable to refrain from sinning. Although we retain the freedom to choose according to our desires, our desires are intrinsically enslaved to corrupt impulses.

Augustine contended that this remaining vestige of freedom inevitably leads to sin, accentuating humanity's wretched state apart from divine grace. Thus, in the flesh we are free only to sin, which is a freedom without liberty, essentially a *moral* bondage, as Alistair McFayden summarizes,

> True liberty can only come from without, from the work of the grace of God on the soul. Our present sinful condition is inherited by each of us before we are ever able to choose it. We are born into lives, relationships, and social structures distorted in all kinds of ways by sin. These are distortions to our beliefs, desires, affections, and wills so that as soon as we are capable of willing and acting ourselves, we do in ways distorted by sin. Thus, we contribute our willed and chosen sins to the condition of sin we have inherited. Original sin does not alter our capacity for willing and choosing, but rather *co-opts and diverts it.*"[54]

In this view, we are responsible or accountable to God for our present state of sinfulness even though we inherited it before we could ever choose it and are unable to avoid it.[55]

From Anxiety to Epigenetics: Rethinking Hereditary Sin with Kierkegaard

Even from a purely philosophical perspective, the Danish theologian and philosopher Søren Kierkegaard (1813–1855), often regarded as the father of existentialism, struggled with the complexities of faith, freedom, and human existence. In his seminal

54. McFadyen, *Bound to Sin*, 16–18.
55. Messer, *Theological Neuroethics*, 96.

text, *The Concept of Anxiety* (1844), subtitled *A Simple Psychologically-Oriented Deliberation in View of the Dogmatic Problem of Hereditary Sin*, Kierkegaard addresses the tension between human freedom and the inherited burden of sin, comparable to Augustine's elaborate descriptions. Kierkegaard's insights on anxiety—a condition he described as the *dizziness of freedom*—provide a profound framework for understanding inherited human frailty,

> Anxiety may be compared with dizziness. He whose eye happens to look down into the yawning abyss becomes dizzy. But what is the reason for this? It is as much in his own eye as in the abyss, for suppose he had not looked down. Thus, anxiety is the dizziness of freedom, which emerges when the spirit would posit the synthesis and freedom looks down into its own possibility, laying hold of finiteness to support itself.[56]

Kierkegaard presents anxiety as more than a psychological condition; it is a spiritual state intrinsically tied to human freedom and the inherited legacy of original sin. Building on Augustine's doctrine, which identifies original sin as a corruption passed down through human nature, Kierkegaard explores how this inheritance creates a tension between our freedom to choose and a propensity toward wrongdoing. For Augustine, this corruption manifests as a *massa peccati*—a mass of sin—leaving humanity unable to fully refrain from sinning, even with free will. Kierkegaard adds that anxiety arises from the recognition of possibility: the capacity for both good and evil. He likens this to standing at the edge of a precipice, gripped by the simultaneous fear of falling and the allure of leaping. This image is profoundly comparable with Augustine's view of human desires enslaved by sin and also explains the nature of addiction. In both cases, freedom becomes obsessed with compulsion—a "dizziness of freedom" that is comparable to the pathological craving inherent in addiction. Augustine's emphasis on the corrupted will, alongside Kierkegaard's existential perspective, provides a powerful method for understanding the spiritual

---

56. Kierkegaard, *Concept of Anxiety*, 75.

and psychological aspects of human struggle, particularly in the context of compulsive behaviors like addiction.

This existential analysis finds a parallel in the scientific revelations of epigenetics. Just as Kierkegaard identified anxiety as an inherited spiritual burden, epigenetics uncovers how environmental experiences—such as trauma or addiction—leave molecular imprints that affect not only individuals but also their descendants.

As research studies conducted by Meaney et al. demonstrate, chronic stress can alter DNA methylation patterns, influencing future generations' susceptibility to anxiety, maladaptive behaviors, and addiction.[57] This biological inheritance can be likened to Kierkegaard's portrayal of hereditary sin—a legacy passed down through generations, influencing human behavior and intensifying the struggle for freedom. Further, Kierkegaard's idea of despair—a condition born of anxiety and the inability to overcome inherited sin—profoundly confirms our understanding of epigenetic constraints.

For instance, children of trauma survivors often exhibit heightened stress responses or predispositions to addiction, not because of a genetic defect but due to inherited epigenetic changes in DNA methylation.[58] These changes, like Kierkegaard's concept of hereditary sin, weigh on individuals, influencing their sense of agency and freedom. Yet, for Kierkegaard, anxiety is not merely a curse, it is also a call to action—a recognition of the need for grace to override inherited limitations. Redemption, in Kierkegaard's mind, lies in one's steadfast faith, which enables individuals to rise above despair and reclaim their freedom. Similarly, the possibility of epigenetic reprogramming can be evidenced where therapeutic interventions, environmental changes, and supportive relationships can disrupt maladaptive gene expression.[59] Just as grace offers a way to break the cycle of sin, epigenetic interventions provide a means to overcome inherited burdens.

57. Meaney, "Environmental Programming of Stress Responses," 1013–24.
58. Miller, "Harsh Family Climate in Early Life," 742–49.
59. Tremblay, "Developmental Origins of Disruptive Behavior Problems," 341–67.

From Augustine's writings on original sin to Kierkegaard's reflections on anxiety, the discussion of hereditary sin bridges theology's exploration of human struggle with our efforts to understand and heal inherited trauma and addiction. Together, they offer the possibility of breaking free from inherited cycles and living a transformed life through both grace and an understanding of epigenetic inheritance.

IMPRINTS OF EXPERIENCE: HOW ENVIRONMENT SHAPES ADDICTION

Understanding the ontological foundation of addiction, what is it about environment that plays such a pivotal role in developing and treating it? For that matter, how does environment influence the terrain or landscape as a predisposition of a particular type of addiction? With these critical questions in mind, it cannot be overstated that environment has significant effects on behavior, particularly a stressful environment and how this can lead to addictive behavior. Stressful circumstances that lead to an adverse environment in early life can force an adolescent to engage in survival strategies that often lead to self-destructive behaviors, such as addiction. Adverse circumstances often lead to attachment disorders as well, which are also associated with a greater likelihood of substance-use problems later in life.

When social interactions rely on mutual struggles with drug, alcohol, or other addictions, it can be nearly impossible to remove oneself from these circumstances and lose the sense of connection. The power of belonging is fueled by the addictions because the relationship is centered on the struggle. Yet when one successfully leaves the group for a more productive, satisfying life, one can often relapse and return to the previous lifestyle. A successful strategic coping mechanism that is at the core of resilience is knowing one's environmental triggers in order to minimize their effects. This provides greater control during the stages of recovery. With this understanding, the risk of relapse can be minimized by integrating an awareness of one's environment with the changeability of epigenetic mechanisms, leading to a balance between the freedom of the will, hope, and resilience.[60]

From the perspectives of theology and epigenetics, the question remains: Why do some drug users continue to seek out drugs, despite the prospect of losing family, friends, health, or livelihood? As epigenetic alterations occur based on changes in an environment, Jean Lud Cadet explores how an early drug-using environment often develops into persistent and powerful triggers for relapse.[61] Epigenetic factors provide an answer to the challenge as to how transient experiences lead to long-lasting risk for relapse in users who try to quit. To determine the motivation to seek out drugs even after long periods of abstinence, it is important to consider the spiritual nature as well as the physical.

The brains of drug users who have progressed to addiction differ markedly from those of early or casual users. Long-lasting associations form between the early use of a drug and different aspects of the early drug-using environment, such as the location in which a drug was first taken or the emotions a user was experiencing at the time. This can cause addicted users who have quit to experience cravings when in a similar setting. This can be a major problem, particularly in the treatment of addiction following a rehabilitation program. As discussed earlier, the endogenous neurosteroid, DHEA (dehydroepiandrosterone), caused a reduction in relapse rates on adult polydrug users while they were taking part in a detoxification program.[62] The study's findings suggest that relapse in addictive behavior can be influenced by epigenetic changes following the administration of DHEA. The details of these epigenetic mechanisms are currently being investigated, as this research may be of considerable value for building greater resilience and reducing the risk of relapse.

60. Hjemdal et al., "Resilience Is a Good Predictor of Hopelessness," 174–80.

61. Cadet, "Epigenetics of Stress," 545–60.

62. Lax et al., "DNA Methylation Signature of Addiction," 322.

GENES OF REDEMPTION: A NEW APPROACH TO ADDICTION RECOVERY

The question of whether a theological approach to treating addictive behavior is better than traditional methods currently used is subjective and depends on individual beliefs, perspectives, and the specific context of the treatment. Both theological and traditional approaches have their merits, and what may work for one person may not work for another. It's important to consider a balanced view when examining this topic. The theological approach views addiction as a multi-dimensional issue, encompassing physical, psychological, social, and spiritual aspects, and individuals may find a more comprehensive path to recovery. By realizing that addiction is not merely a matter of physical dependency, one can develop a greater understanding that involves complex interactions between biological, psychological, social, and spiritual factors. This broader understanding allows for a greater emphasis on the purpose and meaning in life. This can provide individuals with a sense of identity, hope, and a reason to change their addictive behaviors by helping them explore questions of meaning and purpose. Theological approaches can address these underlying existential concerns and provide a sense of hope and belonging.

Communities can provide a strong support network and encouragement from like-minded individuals, which are essential during the recovery process. The power of community in the recovery process cannot be overstated. Addiction often thrives in isolation, and individuals struggling with addictive behaviors may feel ashamed, stigmatized, or disconnected from others. By providing a supportive community of peers, mentors, and spiritual leaders, theological approaches create a space where individuals can share their experiences, receive encouragement and guidance, and feel accepted and valued for who they are. The sense of belonging and connection that comes from being part of a supportive community can be a powerful antidote to the isolation and alienation that often accompany addiction. One of the key contributions to the success of Alcoholics Anonymous is the discipline of attending organizational meetings. A primary focus of the theological approach allows individuals to draw strength from prayer, teaching of God's word and fellowship as a source of comfort, hope, and resilience during difficult times. It is important however, to acknowledge that theological approaches should respect individuals' autonomy and individuality, particularly the challenge in the diversity of religious and spiritual beliefs among individuals seeking treatment.

Ultimately, the effectiveness of a theological approach versus traditional methods depends on the individual's personal beliefs and by integrating theological principles with evidence-based practices, clinicians, and practitioners. Integrating knowledge of epigenetic mechanisms helps us understand an individual's vulnerability to addiction. Environmental factors such as stress and trauma can influence DNA methylation patterns, altering gene expression in brain regions involved in the development and maintenance of addictive behaviors. Epigenetic studies offer the potential to identify biomarkers of addiction recovery that predict treatment outcomes and inform personalized intervention strategies. By profiling DNA methylation patterns in individuals undergoing addiction treatment, researchers can identify epigenetic signatures associated with treatment response, relapse risk, and long-term recovery. These epigenetic biomarkers may serve as objective measures of treatment efficacy and help clinicians tailor interventions to individuals' unique biological and psychosocial profiles. For example, changes in DNA methylation levels in the promoter regions of genes encoding opioid receptors, cannabinoid receptors, and dopamine transporters have been correlated with treatment response to pharmacotherapies such as methadone, buprenorphine, and naltrexone in individuals with opioid use disorder.[63]

As we have seen, however, epigenetics offers a novel framework for understanding the biological, psychological, social, and spiritual dimensions of addiction vulnerability and recovery. By elucidating the mechanisms that occur between

---

63. Kaplan et al., "DNA Epigenetics in Addiction Susceptibility," 2–4.

genes, environments, and experiences, epigenetic research has the potential to revolutionize addiction prevention, treatment, and recovery. By integrating epigenetics into addiction research and practice, we can develop personalized interventions that address individuals' unique biological and psychosocial needs, promote resilience and well-being, and transform lives affected by addiction. Given the description of Augustine's view of original sin and its potential parallels with addictive behavior, we can explore how epigenetic changes in genes related to addictive behavior may be viewed within this framework. Without any evidence that Augustine could even hypothesize a concept such as epigenetics, we can consider how his understanding of inherent human weakness and the struggle for redemption align with our understanding of epigenetic mechanisms and their influence on addictive behavior. In Augustine's view, original sin introduces a fundamental flaw in human nature, resulting in a state of spiritual and moral bondage. As elaborated on previously, his perspective aligns with the concept of addiction as a manifestation of disordered desires and the loss of autonomy.

Epigenetic changes in genes related to addictive behavior can be seen as reflecting the consequences of original sin at a biological level, further contributing to the inherent human weakness and susceptibility to addiction. Augustine's understanding of original sin emphasizes the brokenness and disordered desires in human nature, which can be paralleled with the dysregulation of reward pathways observed in addiction. Understanding that epigenetic modifications are reversible and can be influenced by environmental factors is essential when considering interventions and therapeutic approaches. Just as Augustine emphasizes the importance of grace in redemption, the concept of epigenetic plasticity suggests that interventions, such as behavioral therapy, faith-based community groups, and environmental changes, can modify an individual's methylscape and potentially restore normal gene expression patterns associated with healthy behaviors.

Breaking the Cycle: Redemption's Transformative Power

Christian recovery programs offer vital support to those trapped in addiction, guiding them toward a life centered on redemption. Through this redemptive focus, individuals are empowered to *change one's mind*, breaking the cycle of cravings and adopting healthier coping mechanisms. Liberation from addiction goes beyond mere behavioral change—it involves a profound transformation of a person's moral and spiritual condition, enabled by the power, forgiveness and renewal provided by the Holy Spirit. Faith-based interventions—such as prayer and Scripture, offer an alternative means of calming the chaotic internal environment of the brain. Programs aimed at developing an ARTS (Action, Reaction, Thinking Smart) strategy in adversity help individuals break free from the mental turbulence that exacerbates mental health issues, offering a spiritual complement to conventional care.[64]

From an epigenetic perspective, addiction arises from inherited propensities and environmental influences, paralleling the spiritual inheritance of original sin. By applying the ARTS strategy alongside Augustine's insights into our inherited disposition, an individual can achieve true liberation through what I refer to as the *Cycle of Redemption*. This cycle is a transformative process that counters the Cycle of Addiction, discussed previously, in that it builds enduring faith by a dependance on God, as outlined in 2 Pet 1:5–7: "But also for this very reason, giving all diligence, add to your faith virtue, to virtue knowledge, to knowledge self-control, to self-control perseverance, to perseverance godliness, to godliness brotherly kindness, and to brotherly kindness love." This passage exemplifies the steps toward spiritual growth and freedom, where faith builds virtue, self-control, perseverance, and ultimately, love. Each step of the cycle reflects a deeper transformation, not just of behavior but of the very epigenetic influences that may predispose one to addiction, as presented in Figure 4.

---

64. Varghese, "Role of Spirituality in the Management of Major Depression," 1–2.

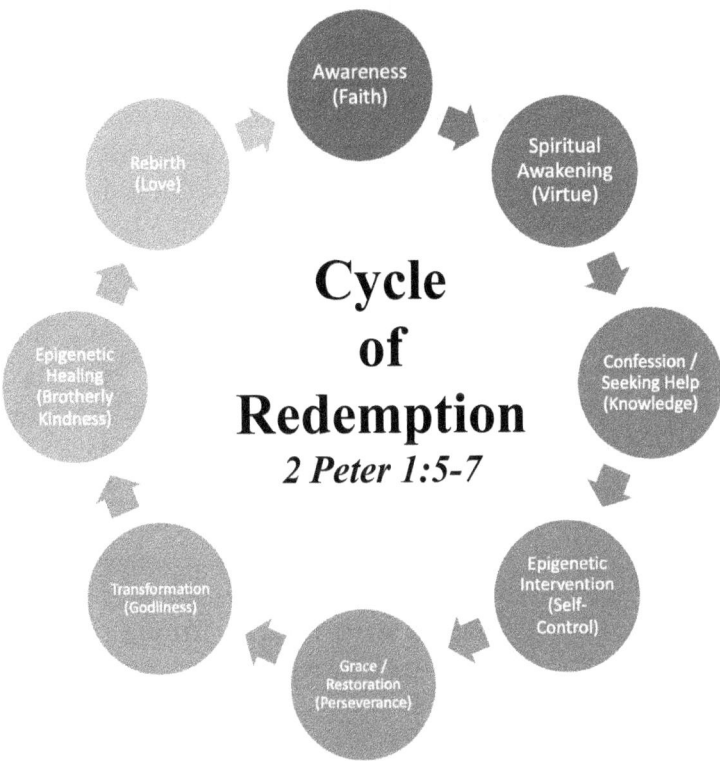

Figure 4: *The Cycle of Redemption*. In 2 Pet 1:5–7, the progression of a life of redemption begins with an awareness of the problem, specifically addiction. The craving is met is with a prompting of hope for overcoming the initial trigger of wanting. The spiritual awakening leads to virtue, integrity, and a commitment to change, while knowledge supports healing through guidance from the Holy Spirit or others. This deepens self-control, supported by epigenetic changes (e.g., in the *BDNF* gene), which reprogram responses to cravings. Grace cultivates perseverance, helping the individual endure challenges, while enabling transformation into godliness. This leads to brotherly kindness, as empathy grows through shared struggles, and culminates in love, evidenced by spiritual maturity and a life of freedom. Each stage of growth builds on the previous, forming an upward cycle of redemption.

The Cycle of Redemption, as described in Figure 4, is based on spiritual growth through the stages detailed in 2 Pet 1:5–7, beginning with an *awareness* of the problem, an initial awakening that prompts seeking help. Faith becomes the foundation of change, building trust in something beyond oneself that provides strength and hope to confront the addiction. This faith anchors the process by guiding the cycle of redemption forward. From here, the process advances with a *spiritual awakening* or *virtue*, reflecting a commitment to integrity and the courage to break free from the addictive cycle. This awakening is a transformative step, illuminating the path toward change. Guided by a heart and mind to reason, a *knowledge* that leads to seeking guidance—whether through the Holy Spirit, a faith community, or therapeutic support. Here, one explores the root causes of struggles and learns constructive healing strategies through Scripture meditation and prayer. This knowledge builds the foundation for *self-control*, where the ability to resist cravings and managing impulses grows, leading to a reprogramming of responses. At this stage, epigenetic changes, such as those derived from the activity in the *BDNF* gene, support adaptive responses to recovery.

With genes and behaviors being re-oriented, the grace of God promotes *perseverance* (steadfastness), an equipping to endure challenges so that the spiritual and emotional progress can move toward restoration. Through faith, and most often forgiveness, and sometimes medical support, resilience can be cultivated. Grace, as an *affordance*, provides the spiritual strength to continue despite obstacles, reflecting a persistent transformation due the "environment" of God's revelation.[65] From perseverance

65. A more elaborate definition of theological affordance, first described by psychologist James Gibson (1979), is the potential for theological insight, understanding, or action that arises from the interaction between an individual's spiritual, cognitive, and moral capacities and the "environment" of divine revelation, Scripture, traditions, or lived experiences of faith. It emphasizes the ways theological concepts, symbols, or practices invite engagement or action—such as prayer, repentance, or worship—depending on the believer's spiritual disposition and situational context. For instance, the sacrament of communion affords spiritual renewal and reflection for those who approach it with faith and understanding, shaped by the theological framework that

springs *godliness* (transformation), where lasting changes in behavior, relationships, and thoughts take root. This godliness grows into *brotherly kindness*, which is a disposition of empathy and attunement to others experiencing similar struggles. During this stage, epigenetic mechanisms may either express or silence genes involved in addictive behavior, rewiring neurological pathways toward healing. Finally, the cycle culminates in *love*, the ultimate expression of spiritual maturity—a *rebirth* that brings Christlike character and liberty. This transformative love is the ultimate expression of godly character. It is a sincere, selfless love that seeks to serve others first. For true freedom is not autonomy, but a self-sacrificial life that acknowledges the reality of sin, embraces continual confession, and seeks restoration of a genuine relationship with God. Each stage builds upon the previous, forming a continuous cycle of progressive sanctification, evidenced by spiritual growth.

By living out the Cycle of Redemption, individuals rewire their patterns of dependency on oneself, both their spiritual and physiological being, through the power of the Holy Spirit, developing an entire dependence on God. With faith as the starting point, it is crucial for one who chooses to move forward in their recovery, to believe that change is possible by the power of the Holy Spirit. That faith serves as the cornerstone of the journey towards recovery from addiction. It involves not only belief in the possibility of change but also trust in the intervention of God. Within the foundation of theology and epigenetics, faith can be understood as the catalyst for transformation by initiating the epigenetic changes in genes, specifically the *BDNF* gene.

Trusting God for his power to renew and transform the heart and mind, as spoken about in Rom 12:2, lays the foundation for the entire cycle. Expanding on the concept of faith, it's essential to cultivate virtue, demonstrated by qualities such as honesty, integrity, and humility, that lead to wise decision-making, so that individuals discern when harmful behaviors conflict with their values. These attributes increase knowledge that serves as a powerful tool in recovery, empowering individuals with self-control to break

---

defines its significance.

habits and manage urges. Self-control is a critical skill in recovery as it enables individuals to resist cravings and impulses by developing awareness of their thoughts, emotions, and sensations. Being persistent in self-control develops perseverance, which is also essential for overcoming the obstacles and setbacks encountered during recovery. Perseverance is the ability to bounce back from adversity, adapt to change, and thrive in the face of challenges and it is the foundation of successful recovery leading to resilience. As this attribute is dependent on the intervention of God, perseverance develops a greater sense of godliness, by cultivating an intimate relationship with God. It is from this true dependance on the power of God that individuals are strengthened and are drawn to others who are experiencing similar challenges—evidenced by brotherly kindness which is at the core of a caring community.

By living in a community with a strong sense of belonging, the supportive network becomes one of the major success factors in lasting recovery, which is one of the primary success factors in programs such as Alcoholics Anonymous (AA) and Narcotics Anonymous (NA). These programs provide individuals with a safe and non-judgmental space to share their experiences, receive encouragement, and learn from others who have walked similar paths resulting in a community of love and forgiveness—which are the lasting attributes defined in the cycle of redemption. By living in a faith-based community that promotes virtue, pursues knowledge, practices self-control, perseveres through challenges, and encourages godliness, brotherly kindness, and love, individuals can confidently proceed through the stages of recovery with lasting resilience that is essential for transformation.

BUILDING RESILIENCE: A KEY STRATEGY FOR RELAPSE PREVENTION

The Brain-Derived Neurotrophic Factor (*BDNF*) gene plays a major role in understanding the interaction between genetic predispositions and environmental influences on both behavior and brain function. BDNF is essential for neuroplasticity, allowing the brain to adapt to new experiences, recover from stress, and learn from environmental inputs. The epigenetic regulation of *BDNF* expression—influenced by factors such as trauma, nutrition, exercise, and even spiritual practices—reflects the complex interaction between human nature and life's circumstances. While individuals are born with certain genetic predispositions, their capacity for change is shaped by their environment. This concept is the foundation of previous theological discussions on the dual forces of nature and grace in human development, specifically Augustine's reflections on original sin. In this context, BDNF serves as a biological counterpart to these theological ideas, suggesting that while humans are predisposed to certain vulnerabilities, they also possess the potential for healing and change.

The role of BDNF in addiction research is particularly significant, earning the gene its title as the "resilience gene." Research has shown that reduced *BDNF* expression is linked to impaired neuroplasticity, leaving individuals more susceptible to addictive behaviors and less able to develop healthy coping strategies. Chronic substance abuse, trauma, and sustained environmental stress can further suppress *BDNF* expression, perpetuating a cycle of dependency. This biological vulnerability is reminiscent of Augustine's theological assertion that sin corrupts human nature, causing individuals to struggle with the tension between free will and the compulsion toward wrongdoing. Søren Kierkegaard's concept of anxiety as a "dizziness of freedom" provides a helpful perspective for understanding this internal conflict, illustrating the role of the conscience as it struggles with the freedom to choose but also the pull of negative, habitual behaviors.

In response to this dilemma, research supports the healing potential of spiritual practices such as prayer, meditation, and discipleship, which have been shown to regulate stress responses and lower cortisol levels—biological markers of stress.[66] These practices have also been found to upregulate *BDNF* expression, enabling neuroplasticity and contributing to recovery. Studies in *psychoneuroimmunology* (the study of the effect of the mind on

---

66. Zablocki et al., "Spiritual Practices in Addiction Recovery," 14–19.

health and resistance to disease) emphasize the connection between stress reduction and increased BDNF levels, offering strong support for the integration of spiritual practices as complementary therapeutic tools.[67] Such programs not only address the psychological and moral aspects of addiction, but also create an environment of purpose and belonging—key elements of recovery. By nurturing positive emotional states and building a supportive community, these programs serve as a form of *epigenetic intervention*, affecting both the spiritual and biological aspects of addiction. This supportive environment allows individuals to form healthier neural pathways, bridging genetic predispositions to addiction with the transformative potential for recovery.

Resilience training plays a pivotal role in addiction recovery, equipping individuals with the skills needed to manage stress, resist cravings, and maintain long-term recovery. Research by Witkiewitz and Marlatt emphasizes the importance of emotional regulation, problem-solving abilities, and self-awareness in preventing relapse.[68] Resilience encompasses not only the ability to recover from adversity but also the ability to thrive in the face of challenges. Developing resilience requires deliberate skill-building, such as improving stress management, adaptive thinking, and creating a strong network of social support. While socioeconomic factors influence resilience, close personal relationships—particularly with family and friends—are crucial in encouraging consistent, positive choices. When individuals face moments of vulnerability, resilience training provides a toolkit of coping mechanisms, such as cognitive reframing and problem-solving. These help individuals resist the temptation to relapse and renew their commitment to personal goals. Faith-based programs further enhance these resilience-building efforts by integrating spiritual discipline with practical strategies for recovery. These programs emphasize accountability, community support, and a sense of purpose, all of which strengthen an individual's ability

---

67. Kiecolt-Glaser et al., "Behavioral Influences on Immune Function," 991–99.

68. Witkiewitz et al., "Relapse Prevention for Alcohol and Drug Problems," 224–35.

to stay focused on long-term recovery. The integration of theological insights with evidence-based practices offers a systematic approach to relapse prevention, addressing both the psychological and biological dimensions of addiction while promoting the personal growth necessary for lasting recovery.

Unshackling the Soul: Epigenetics, Original Sin, and the Power of the Spirit

C. S. Lewis once wrote, "Reason is the organ of truth, imagination is the organ of meaning." This insight provides a fitting way to reflect on what *Genes of Eden* seeks to achieve. The science of epigenetics offers us truths about how addiction, trauma, and inherited behaviors are etched into human biology, revealing the intricate workings of our fallen condition. Epigenetic changes, such as those caused by early life adversity, addiction, or trauma, disrupt biological processes, shaping behaviors and binding individuals to cycles of struggle. Augustine's doctrine of original sin, while not addressing epigenetic mechanisms, provides a perspective through which to understand the universality of human fallibility and its consequences. His reflections on the tension between human will and divine grace demonstrate the disrupted biological processes seen in addiction, where compulsive behaviors are evidenced by a spiritual battle as Augustine described. Epigenetics, in turn, deepens our understanding of how these struggles can transcend generations, leaving a biological imprint of humanity's propensity for sin. Yet, both science and theology offer more than diagnosis—they provide affordances, opportunities for healing and transformation.

The cycle of addiction, with its relentless grip on the body, mind, and soul, finds its contrast in the cycle of redemption offered through Christ. Salvation through the grace of Jesus Christ and the indwelling of the Holy Spirit provides freedom from sin's grip, not just spiritually but also in the body and brain. The eighth chapter of Romans reminds us that the Holy Spirit, who raised Christ from the dead, lives within every believer, empowering them with the resilience to overcome compulsive desires and walk in newness of life. Through the Holy Spirit's power, the brain-body-soul connection can be reoriented toward peace, opening pathways of neuroplasticity and healing. For those whose methylscapes bear the scars of early life trauma, resilience becomes possible—not by sheer willpower but through a life yielded to God. This is the true essence of soul training: engaging the mind, body, and spirit in a continuous walk with God, fortified by Scripture, prayer, and community. To further support soul training, studies conducted by Dvorak et al. also emphasize that personal responsibility enhances self-regulation as a tool in reducing vulnerability to addiction.[69]

Reason and imagination together illuminate the path forward. Science provides the truth of how addiction reshapes us, while theology gives meaning to the redemptive possibilities that lie beyond the brokenness. *Genes of Eden* calls us to see that humanity is not merely bound by the past but capable of transformation—through grace, responsibility, and the ever-present work of the Holy Spirit. In this convergence of truth and meaning, the cycles of addiction and redemption stand as vivid reminders of both human weakness and God's enduring love and hope. By understanding these cycles, we gain the strength to seek true freedom—not just for ourselves, but for those we love and for the generations that follow.

---

69. Dvorak et al., "Interactive Effect of Impulsivity-Like Traits," 646–54.

Epilogue

Quantum Epigenetics and the Path of Sanctification

IN CONTEMPLATING THE MESSAGES from *Genes of Eden*, particularly the theories of the soul and sin and how they are intertwined into the human genome, there seems to emerge, although speculative, the concept of *quantum epigenetics*. By extension of John Polkinghorne's seminal book, *Quantum Mechanics and Theology: An Unexpected Kinship*,[1] an intriguing possibility lies in a quantum model wherein original sin may be understood as "encoded" within the epigenome, subject to potential quantum influences. In this view, inherited sin could be stored within the methylscape, patterns of which may be influenced by quantum-level phenomena, such as quantum *entanglement* or *decoherence*. This model opens avenues for understanding how deeply ingrained human dispositions, beyond classical genetic inheritance, might be transmitted and modified, not just through direct molecular processes, but possibly through quantum effects that can shape epigenetic expression across generations. Such an approach hints at an intersection of theology, epigenetics, and quantum biology, proposing that our inherited predispositions may carry a complexity that reaches beyond molecular mechanics into the quantum domain. This perspective suggests that original sin could be inscribed into the human genome at a quantum level, influencing the epigenome in a way that predisposes all individuals to moral imperfection or

---

1. Polkinghorne, *Quantum Mechanics and Theology*, ch. 2.

separation from God. By means of sanctification, however, this altered nature presumably becomes maleable as a believer develops a greater tendency towards the holiness of God.

At the intersection of quantum mechanics and epigenetic processes, such as DNA methylation, quantum epigenetics could offer a plausible way to conceptualize the nature of sin embedded within the genome, as well as the mechanisms involved in sanctification. As a believer undergoes spiritual transformation, becoming sanctified, more aligned with the nature of Jesus through the work of the Holy Spirit, could it possible to conceive that this is evidenced in the human epigenome itself? Conceptually, the speculation of quantum epigenetics invites us to consider that sanctification might affect not just the soul but also the body, potentially influencing how our genes are expressed in new ways.

The attributes of quantum mechanics can be observed at the core of quantum epigenetics, such as *interconnectedness*, or *quantum entanglement*, in which the mysterious phenomenon where particles, once linked, remain intertwined no matter the distance. If we consider sanctification through this image, we might imagine a process where the Spirit's influence permeates through the believer's entire being, affecting not just thoughts and actions, but the very biological processes that shape consciousness. Another attribute is the quantum idea of *non-locality*, that interactions can transcend time and space, and align with the concept of God's omnipresence in the sanctifying work of the Holy Spirit, who is unbounded by physical constraints. In essence, the Holy Spirit's presence can touch every corner of a person's life and reach into the molecular world of DNA, shifting the methylscape as it adapts to the stress response, emotional regulation, and even the inclination towards holiness.

Sanctification has always been seen as a journey of becoming—a process where the believer, through grace, gradually conforms to the image of Christ. Quantum mechanics introduces another concept referred to as *wave function*, which is a field of possibility that collapses into reality upon observation. Perhaps the journey of sanctification can be seen in similar terms of a wave in which a believer's potential for goodness, love, and compassion *collapses* into tangible expressions through which choices are made under God's guidance. The concept of *coherence* is another attribute found in quantum mechanics, and it refers to a state where quantum particles act in harmony, maintaining a unified behavior—until interactions with their environment cause a loss of this harmony, known as *decoherence*. Similarly, in the spiritual process of sanctification, a believer strives for a state of alignment, where the heart, mind, and actions are unified with God's will. If we think of sanctification as achieving this kind of spiritual coherence, quantum epigenetics might offer a way to envision how such alignment could have tangible biological effects. For example, spiritual growth could have a stabilizing effect on the biological environment of certain genes, especially by the actions of the *BDNF* gene, which plays a key role in neuronal stability that could result in spiritual coherence. This state of peace achieved through spiritual practices like prayer, may influence the methylscape around the *BDNF* gene leading to a more stable methylation pattern. This will enhance the expression of the *BDNF* gene to support mental and emotional resilience, helping individuals remain strong when encountering the stresses at the root of addiction.

The exploration into the nexus of epigenetics, original sin and addiction, reveals that quantum epigenetics could offer a model to understand sanctification as a process that transcends into the deep biological realities that shape us. It suggests a world where the grace that reshapes a believer's heart might also tune the rhythms of their genes, as indicated by Paul in his conclusion to his first letter to the Thessalonians, "Now may the God of peace himself sanctify you completely and may your whole spirit and soul and body be kept blameless at the coming of our Lord Jesus Christ" (1 Thess 5:23).

While quantum epigenetics remains a speculative bridge between science and theology, it invites us to envision sanctification not as a purely spiritual transformation but as a spiritual and somatic reality, where the sacred and the material, the spiritual and the biological, converge in each transformed soul.

In agreement with Amos Yong's passion to "engage theological authenticity in the encounter with science," and "to present our ideas self-critically, provisionally, and dialogically," I have sought to limit the metaphysical with current, robust science in the hope of attaining genuine truth. This book has sought to bring clarity to Yong's hypothesis:

> the many tongues of Pentecost can, in our time, be understood to include the diverse languages of modern science, each bearing witness in its own way to the work of the Spirit, who is heralding and ushering the Kingdom of God "in the last days".[2]

Through his hypothesis, quantum epigenetics and other scientific perspectives offer a language through which the enigma of transformative spiritual regeneration is reflected in the complexities of human biology and, potentially, the quantum processes that underlie our very nature.

---

2. Yong, "How Does God Do What God Does?," 52.

Bibliography

Allen, Paul L. "Augustine and the Systematic Theology of Origins." In *Augustine and Science*, edited by John Doody et al., 1–20. Lanham, MD: Lexington, 2013.
———. *Ernan McMullin and Critical Realism in the Science-Theology Dialogue*. London: Routledge, 2016.
Aquinas, Thomas. "Nature of the Estimative Sense." *Summa Theologiae* I, 78. Notre Dame, IN, 2024. https://www.scribd.com/document/621907384/Conceptual-Knowledge-Aquinas.
———. "The Union of the Soul with the Body." *Summa Theologiae*, I, 76. Notre Dame, IN, 2024. https://www3.nd.edu/~afreddos/summa-translation/Part%201/st1-ques76.pdf.
Augustine. "Answer to Julian: Answer to the Pelagians." In *The Works of Saint Augustine*, edited by John E. Rotelle, OSA, and translated by Roland J. Teske, 623–61. Hyde Park, NY: New City, 1998.
———. "Augustine Confesses that He Had Formerly Been in Error Concerning the Grace of God." In *Treatise on the Predestination of the Saints*, edited by Philip Schaff and Benjamin Breckinridge Warfield and translated by Robert Ernest Wallis and Peter Holmes, 1–4. https://www.newadvent.org/fathers/1512.htm.
———. *Confessions*. Translated by John K. Ryan. New York: Doubleday, 1960.
———. *Concerning the City of God Against the Pagans*. Translated by Henry Bettenson. London: Penguin, 1984.
———. *The Literal Meaning of Genesis*. Translated by John Hammond Taylor. New York: Newman, 1982.
———. "Marriage and Desire. Answer to the Pelagians." In *The Works of Saint Augustine*, edited by John E. Rotelle, OSA, and translated by Roland J. Teske, 418–21. Hyde Park, NY: New City, 1998.
———. "The Nature and Origin of the Soul." In *The Works of Saint Augustine*, edited by John E. Rotelle, OSA, and translated by Roland J. Teske, I.12.15. Hyde Park, NY: New City, 1997.

———. "On the Trinity." In *Nicene and Post-Nicene Fathers*, edited by Paul A. Boer and translated by Arthur West Haddan, 1:3. Tyler, TX: Veritatis Splendor, 2012.

———. "The Perfection of Human Righteousness." In *The Works of Saint Augustine*, edited by John E. Rotelle, OSA, and translated by Roland J. Teske, 13.31; 21.44. Hyde Park, NY: New City, 1997.

———. "The Punishment and Forgiveness of Sins and the Baptism of Little Ones: Answer to the Pelagians I." In *The Works of Saint Augustine*, edited by John E. Rotelle, OSA, and translated by Roland J. Teske, SJ, 411–12. Hyde Park, NY: New City, 1997.

———. "Unfinished Work in Answer to Julian. Answer to the Pelagians, III." *The Works of Saint Augustine*, edited by John E. Rotelle, OSA, and translated by Roland J. Teske, SJ, 623–61. Hyde Park, NY: New City, 1999.

Babcock, William. "The Ethics of St. Augustine." *Journal of Religious Ethics: Studies in Religion* 3 (1992) 116.

Binder, Nadine K., et al. "Parental Diet-Induced Obesity Leads to Retarded Early Mouse Embryo Development and Altered Carbohydrate Utilization by the Blastocyst." *Reproduction, Fertility and Development* 24 (2012) 804–12.

Bohacek, Johannes, et al. "Epigenetic Inheritance of Disease and Disease Risk." *Neuropsychopharmacology* 38 (2012) 220–36.

Boucher, Jean, et al. "Insulin and Insulin-Like Growth Factor 1 Receptors are Required for Normal Expression of Imprinted Genes." *Proceeding in the National Academy of Sciences*, 40 (2014) 14512–17.

Bray, Gerald. *Augustine on the Christian Life: Transformed by the Power of God*. Wheaton, IL: Crossway, 2015.

Cacioppo, John T., et al. "The Bisected Brain." *Perspectives on Psychological Science* 5 (1997) 5–6.

Cadet, Jean Lud. "Epigenetics of Stress, Addiction, and Resilience: Therapeutic Implications." *Molecular Neurobiology* 53 (2016) 545–60.

Cadet, Jean Lud, et al. "Transcriptional and Epigenetic Substrates of Methamphetamine Addiction and Withdrawal: Evidence From a Long-Access Self-Administration Model in the Rat." *Molecular Neurobiology* 51 (2015) 696–717.

Campbell, Robert R., et al. "How the Epigenome Integrates Information and Reshapes the Synapse." *Nature Reviews Neuroscience* 20 (2019) 133–47.

Catechism of the Catholic Church. "§366." 2024. https://www.catholiccrossreference.online/catechism/#!/search/366.

Champagne, Francis A. "Transgenerational Effects of Social Environment on Variations in Maternal Care and Behavioral Response to Novelty." *Behavioral Neuroscience* 122 (2008) 266.

Chesterton, Gilbert Keith. *Orthodoxy*. Milwaukee: Cavalier, 2015.

Chrysostom, John. *Homilies on Genesis, 1–17*. Translated by R. C. Hill. The Catholic University of America, 1986.

BIBLIOGRAPHY

Clark, Tanya T., et al. "Everyday Discrimination and Mood and Substance Use Disorders: A Latent Profile Analysis with African Americans and Caribbean Blacks." *Addictive Behaviors* 40 (2014) 119–25.

Cloud, John. "Why Your DNA Isn't Your Destiny." *Time Magazine* 6 (2010) para. 6. https://content.time.com/time/subscriber/article/0,33009,1952313-8,00.html.

Cook, Christopher H. *Alcohol, Addiction and Christian Ethics.* Cambridge: Cambridge University Press, 2016.

Cortes, Laura R., et al. "Does Gender Leave an Epigenetic Imprint on the Brain?" *Frontiers in Neuroscience* 13 (2019) 1–8.

Couenhoven, Jesse. "St. Augustine's Doctrine of Original Sin." *Augustinian Studies* 36 (2005) 359–96.

———. *Stricken by Sin, Cured by Christ: Agency, Necessity, and Culpability in Augustinian Theology.* New York: Oxford University Press, 2013.

Cunliffe, Vincent T. "The Epigenetic Impacts of Social Stress: How Does Social Adversity Become Biologically Embedded?" *Epigenomics* 8 (2016) 1653–69.

Danielou, Jean. *Origen.* Translated by Walter Mitchell. New York: Sheed and Ward, 1955.

Dias, Brian G., and Kerry J. Ressler. "Parental Olfactory Experience Influences Behavior and Neural Structure in Subsequent Generations." *Nature Neuroscience* 17 (2014) 89–96.

Dietz, David M., et al. "Paternal Transmission of Stress-Induced Pathologies." *Biological Psychiatry* 70 (2011) 408–14.

Doody, John, et al. *Augustine and Science.* Lanham, MD: Lexington, 2013.

Dunnington, Keith. *Addiction and Virtue—Beyond the Models of Disease and Choice.* Downers Grove, IL: InterVarsity Academic, 2011.

Dvorak, Robert D., et al. "The Interactive Effect of Impulsivity-Like Traits and Perceived Responsibility for Alcohol- and Nicotine-Related Consequences." *The American Journal of Drug and Alcohol Abuse* 43 (2017) 646–54.

Earley, Paul H. "America's Addiction Crisis, Compounded by Covid-19, Requires Immediate Action to Save Lives." *Stat News*, February 10, 2021. https://www.statnews.com/2021/02/10/addiction-crisis-covid-19-requires-action.

Edqvist, Per-Hendrik D., et al. "Expression of Human Skin-Specific Genes Defined by Transcriptomics and Antibody-Based Profiling." *Journal of Histochemistry and Cytochemistry* 63 (2015) 129–41.

Erickson, Millard J. *Christian Theology.* Grand Rapids: Baker Academic, 2013.

Franklin, Tamara B., et al. "Epigenetic Transmission of the Impact of Early Stress Across Generations." *Biological Psychiatry* 68 (2010) 408–15.

Fullston, Thomas, et al. "Diet-Induced Paternal Obesity in the Absence of Diabetes Diminishes the Reproductive Health of Two Subsequent Generations of Mice." *Human Reproduction* 27 (2012) 1391–400.

Gaul, Brett. "Augustine on the Virtues of the Pagans." *Augustinian Studies* 40 (2009) 242.

Gage, Shane H., et al. "Rat Park: How a Rat Paradise Changed the Narrative of Addiction." *Addiction* 114 (2018) 917–22.

Geschwind, Nicole, et al. "Meeting Risk with Resilience: High Daily Life Reward Experience Preserves Mental Health." *Acta Psychiatry Scandanavia* 1222 (2010) 129–38.

Gibson, James J. *The Ecological Approach to Visual Perception*. Boston: Houghton Mifflin, 1979.

Goelet, Peter, et al. "The Long and the Short of Long-Term Memory—A Molecular Framework." *Nature* 322 (1986) 419–22.

Goldman, David, et al. "The Genetics of Addictions: Uncovering the Genes." *Nature Reviews Genetics* 6 (2005) 521–32.

Grandjean, Valerie, et al. "RNA: a Possible Contributor to the 'Missing Heritability'. *Basic and Clinical Andrology* 23 (2013) 1–7.

Grant, Robert M. *God and Reason in the Middle Ages*. Cambridge: Cambridge University Press, 2001.

Greer, Rowan A. *Christian Life and Christian Hope: Raids on the Inarticulate*. New York: Crossroad, 2001.

Griffith-Thomas, William H. *Romans: A Devotional Commentary*. London: Religious Tract Society, 1911.

Guerrero-Bosagna, Carlos, et al. "Epigenetic Transgenerational Actions of Vinclozolin on Promoter Regions of the Sperm Epigenome." *PLoS One*. 5 (2010) e13100.

Haidt, Jonathan. *The Righteous Mind: Why Good People are Divided by Politics and Religion*. New York: Vintage, 2012.

Heijmans, Bastiaan, et al. "Persistent Epigenetic Differences Associated with Prenatal Exposure to Famine in Humans." *Proceedings in the National Academy Sciences* 105 (2008) 17046–49.

Heller, Elizabeth A., et al. "The Neuroepigenetics of Addiction." *American Journal of Psychiatry* 173 (2016) 1140–49.

Hjemdal, Odin, et al. "Resilience Is a Good Predictor of Hopelessness Even After Accounting for Stressful Life Events, Mood, and Personality (NEO-PI-R)." *Scandinavian Journal of Psychology* 53 (2012) 174–80.

Hope, Bruce T. "Trans-Generational Resilience to Addiction: Role of Cortical BDNF." *Nature Medicine* 19 (2013) 136–37.

Inagaki, Tristen K. "Opioids and Social Connection." *Current Directions in Psychological Science* 27 (2018) 85-90.

Irenaeus of Lyons. "Against Heresies." In *The Apostolic Fathers with Justin Martyr and Irenaeus*, edited by Alexander Roberts et al., 442–48. Buffalo, NY: Christian Literature Company, 1885.

Jacobs, Alan. *Original Sin: A Cultural History*. New York: HarperCollins, 2009.

Jacqueline, Maryse V. "Genetics of Addiction: Future Focus on Gene-Environment Interaction?" *Journal of Studies on Alcohol and Drugs* 77 (2016) 684–87.

Jiang, Weiquan, et al. "ZFP57 Dictates Allelic Expression Switch of Target Imprinted Genes." *Proceedings of the National Academy of Sciences of the United States of America* 118 (2021) e2005377118.

Johnson, Aubrey R. *The Vitality of the Individual in the Thought of Ancient Israel*. Cardiff: University of Wales Press, 1964.

Kaplan, Graham, et al. "DNA Epigenetics in Addiction Susceptibility." *Frontiers in Genetics* 13 (2022) 1-4.

Keller, Timothy. "How Sin Makes Us Addicts." Sermon, Redeemer Presbyterian Church, New York City, February 16, 2010.

Kellermann, Natan. "Epigenetic Transmission of Holocaust Trauma: Can Nightmares Be Inherited?" *Israel Journal of Psychiatry and Related Sciences* 50 (2013) 33-9.

Kiecolt-Glaser, Janice K., et al. "Behavioral Influences on Immune Function and Disease." *The Journal of Clinical Investigation* 110 (2002) 991-99.

Kierkegaard, Søren. *The Concept of Anxiety: A Simple Psychologically Oriented Deliberation in View of the Dogmatic Problem of Hereditary Sin*. Translated by Alastair Hannay. London: Penguin, 2014.

Klengel, Torsten, et al. "The Role of DNA Methylation in Stress-Related Psychiatric Disorders." *Neuropharmacology* 80 (2014) 115-32.

Klima, Gyula. "Aquinas on the Materiality of the Human Soul and Immateriality of the Human Intellect." *Philosophical Investigations* 32 (2009) 163-82.

Koob, George F. "Neurobiology of Addiction: A Neuro-Circuitry Analysis." *Lancet Psychiatry* 3 (2016) 760-73.

Krentzman, Amy R. "Gratitude, Abstinence, and Alcohol Use Disorders: Report of a Preliminary Finding." *Journal of Substance Abuse and Treatment* 78 (2017) 30-36.

Kubota, Takeo, et al. "Epigenetic Understanding of Gene-Environment Interactions in Psychiatric Disorders: a New Concept of Clinical Genetics." *Clinical Epigenetics* 4 (2012) 1.

Kung, Hans. *Great Christian Thinkers*. New York: The Continuum International, 1995.

Laney, Alexandra et al. "Part 1: What Is Transgenerational Epigenetic Inheritance?" Biology Fortified Inc., 2020. https://biofortified.org/2020/01/transgenerational-inheritance/.

Lax, Etan, et al. "A DNA Methylation Signature of Addiction in T Cells and Its Reversal with DHEA Intervention." *Frontiers in Molecular Neuroscience* 11 (2018) 322.

Leshner, Alan. "Addiction Is a Brain Disease, and It Matters." *Science* 278 (1997) 45-47.

Levine, Amir, et al. "Epigenetic Mechanisms of Addiction: Focus on the BDNF and H3K9me2 Networks." *Genes, Brain and Behavior* 10 (2011) 424-35.

Lilienfeld, Scott O. "Psychological Treatments That Cause Harm." *Perspectives on Psychological Science* 2 (2007) 53-70.

Luce, Patrick. "URI Professor's Study Seeks Better Treatment for Adolescent Addiction." *The Westerly Sun*, May 23, 2021. https://www.thewesterlysun.

com/news/westerly/uri-professor-s-study-seeks-better-treatment-for-adolescent-addiction/article_43e06874-bb92-11eb-a0fd-bf58b6bdcad6.html.

Manikkam, Mohan, et al. "Dioxin (TCDD) Induces Epigenetic Transgenerational Inheritance of Adult-Onset Disease and Sperm Epimutations." *PLoS One* 7 (2012) e46249.

Marshall, Paul, et al. "Cognitive Neuroepigenetics: The Next Evolution in Our Understanding of the Molecular Mechanisms Underlying Learning and Memory?" *Nature Partner Journal: Science of Learning* 1 (2016) 16004.

McFadyen, Alistair. *Bound to Sin: Abuse, Holocaust and the Christian Doctrine of Sin.* Cambridge: Cambridge University Press, 2000.

McGowan, Patrick O., et al. "Broad Epigenetic Signature of Maternal Care in the Brain of Adult Rats." *PLoS One* 6 (2011) e14739.

———. "Epigenetic Regulation of the Glucocorticoid Receptor in Human Brain Associates with Childhood Abuse. *Nature Neuroscience* 12 (2009) 342–48.

McGrath, Alistair E. *Christian Theology: An Introduction.* Hoboken, NJ: Wiley-Blackwell, 2011.

Meaney, Michael J., et al. "Environmental Programming of Stress Responses Through DNA Methylation: Life at the Interface Between a Dynamic Environment and a Fixed Genome." *Nature Reviews Neuroscience* 6 (2005) 1013–24.

Messer, Neil. *Theological Neuroethics.* New York: Bloomsbury T. & T. Clark, 2017.

Miao, Zhuang. "The Relationship Between Stress, Mental Disorders, and Epigenetic Regulation of BDNF." *International Journal of Molecular Sciences* 21 (2020) 1375.

Miller, Gregory. E., et al. "Harsh Family Climate in Early Life Presages the Emergence of a Proinflammatory Phenotype in Adolescence." *Psychological Science* 21 (2010) 742–49.

Mosqueiro, Bruno Paz, et al. "Increased Levels of Brain-Derived Neurotopic Factor are Associated with High Intrinsic Religiosity Among Depressed Inpatients." *Frontiers in Psychiatry* 10 (2019) 671.

Mychasiuk, Richelle, et al. "Parental Enrichment and Offspring Development: Modifications to Brain, Behavior and the Epigenome." *Behavior and Brain Research* 228 (2012) 294–98.

Nestler, Eric J. "Epigenetic Mechanisms of Drug Addiction." *Neuropharmacology* 76 (2014) 259–68.

———. "Epigenetic Mechanisms in Psychiatry." *Biological Psychiatry* 76 (2014) e13–e14.

Niebuhr, Reinhold. *The Nature and Destiny of Man.* New York: C. Scribner's Sons, 1943.

Nielsen, David A., et al. "Epigenetics of Drug Abuse: Predisposition or Response." *Pharmacogenomics* 13 (2012) 1149–60.

BIBLIOGRAPHY

Oakley, Robert H., et al. "The Biology of the Glucocorticoid Receptor: New Signaling Mechanisms in Health and Disease." *Journal of Allergy and Clinical Immunology* 132 (2013) 1033–44.

Oberlander, Timothy F., et al. "Prenatal Exposure to Maternal Depression, Neonatal Methylation of Human Glucocorticoid Receptor Gene (NR3C1) and Infant Cortisol Stress Responses." *Epigenetics* 3 (2008) 97–106.

Origen. *On First Principles*. Translated by G. W. Butterworth. Notre Dame, IN: Christian Classics, 2013.

Pandey, Ghanshyam N., et al. "Brain-Derived Neurotropic Factor and Tyrosine Kinase B Receptor Signaling in Post-Mortem Brain of Teenage Suicide Victims." *International Journal of Neuropsychopharmacology* 11 (2008) 1047–61.

Panksepp, Jaak, et al. "What Is Basic about Basic Emotions? Lasting Lessons from Affective Neuroscience." *Emotion Review* 3 (2011) 387–96.

Pelikan, Jaroslav. "The Christian Tradition: A History of the Development of Doctrine." In *The Growth of Medieval Theology (600–1300)*, 43:2–62. Chicago: University of Chicago Press, 1978.

———. *The Shape of Death: Life, Death and Immortality in the Early Fathers*. New York: Abingdon, 1961.

Pembrey, Marcus E., et al. "Sex-Specific, Male-Line Transgenerational Responses in Humans." *European Journal of Human Genetics* 14 (2006) 159–66.

Philibert, Robert A., et al. "Serotonin Transporter mRNA Levels Are Associated with the Methylation of an Upstream CpG Island." *American Journal of Medical Genetics* 147 (2008) 844–49.

Plantinga, Alvin. "Science: Augustinian or Duhemian?" In *Augustine and Science*, edited by John Doody et al., 153–80. Lanham, MD: Lexington, 2013.

Polkinghorne, John. *Quantum Mechanics and Theology: An Unexpected Kinship*. New Haven, CT: Yale University Press, 2007.

Qiu, Jane. "Epigenetics: Unfinished Symphony." *Nature* 441 (2006) 143–48.

Ramchandani, Shyam, et al. "DNA Methylation Is a Reversible Biological Signal." *Proceedings in the National Academy of Sciences* 96 (1999) 6107–12.

Rasmussen, Kasper Dindler, et al. "Role of TET Enzymes in DNA Methylation, Development, and Cancer." *Genes and Development* 30 (2016) 733–50.

Richmond, Patrick. "An Augustinian Perspective on Creation and Evolution." In *Augustine and Science*, edited by John Doody et al., 181–94. Lanham, MD: Lexington, 2013.

Robertson, Keith D. "DNA Methylation in Health and Disease." *Nature Review in Genetics* 1 (2005) 11–19.

Roohani, Yusuf, et al. "Predicting Transcriptional Outcomes of Novel Multigene Perturbations with GEARS (Geometric Deep Learning Model for Transcriptional Outcomes)." *Nature Biotechnology* 42 (2024) 927–35.

Roth, Tania L., et al. "Lasting Epigenetic Influence of Early-Life Adversity on the *BDNF* Gene." *Biological Psychiatry* 65 (2009) 760–69.
Ruden, Ronald A. *The Craving Brain*. New York: Harper Collins, 1997.
Satel, Sally, et al. "Addiction and the Brain-Disease Fallacy." *Frontiers in Psychiatry* 4 (2013) 1–41.
Sha, Ky. "A Mechanistic View of Genomic Imprinting." *Annual Review of Genomics and Human Genetics* 9 (2008) 197–216.
Sina, Abu Ali Ibn, et al. "Epigenetically Reprogrammed Methylation Landscape Drives the DNA Self-Assembly and Serves as a Universal Cancer Biomarker." *Nature Communications* 19 (2018) 4915.
Skinner, Michael K. "Role of Epigenetics in Developmental Biology and Transgenerational Inheritance." *Birth Defects Research Part C: Embryo Today* 93 (2011) 51–55.
Skinner, Michael K., et al. "Epigenetic Transgenerational Actions of Environmental Factors in Disease Etiology." *Trends in Endocrinology and Metabolism* 21 (2010) 214–22.
Skov-Jeppesen, Sune Moeller, et al. "Changing Smoking Behavior and Epigenetics: A Longitudinal Study of 4,432 Individuals from the General Population." *Chest* 163 (2023) 1565–75.
Spiegel, James S. "Augustine, Evolution, and Scientific Methodology." In *Augustine and Science*, edited by John Doody et al., 195–210. Lanham, MD: Lexington, 2013.
Szyf, Moshe, et al. "DNA Methylation: A Mechanism for Embedding Early Life Experiences in the Genome." *Childhood Development* 84 (2013) 49–57.
———. "The Social Environment and the Epigenome." *Environmental and Molecular Mutagenesis* 49 (2008) 46–60.
Tanner, Norman P. *Decrees of the Ecumenical Councils, Nicaea I to Lateran V*. Washington, DC: Georgetown University, 1990.
Tertullian. "De Anima." In *Ante-Nicene Fathers*, edited by Alexander Roberts and James Donaldson and translated by Peter Holmes, 417–30. Edinburgh: T. & T. Clark, 1885.
———. "Chapter 40." In *Treatise on the Soul*. Kevin Knight: New Advent Group, 2024. https://www.newadvent.org/fathers/0310.htm.
Tremblay, Richard E. "Developmental Origins of Disruptive Behavior Problems: the 'Original Sin' Hypothesis, Epigenetics and Their Consequences for Prevention." *Journal of Child Psychology and Psychiatry* 51 (2010) 341–67.
Trifu, Sorin, et al. "Aggressive Behavior in Psychiatric Patients in Relation to Hormonal Imbalance." *Experimental Therapeutics and Medicine* 20 (2020) 3483–87.
Tuesta, L. M., et al. "Mechanisms of epigenetic memory and addiction." *European Molecular Biology Organization Journal* 33 (2014) 1091–103.
Varghese, Sebastian Paul, et al. "Role of Spirituality in the Management of Major Depression and Stress-Related Disorders." *Chronic Stress* 5 (2021) 1–2.

Vassoler, Fabiana Maria, et al. "The Impact of Exposure to Addictive Drugs on Future Generations: Physiological and Behavioral Effects." *Neuropharmacology* 76 (2014) 269–75.

———. "Mechanisms of Transgenerational Inheritance of Addictive-Like Behaviors." *Neuroscience* 264 (2014) 198–206.

Volkow, Nora Denise, et al. "Relationship Between Subjective Effects of Cocaine and Dopamine Transporter Occupancy." *Nature* 386 (1997) 827–30.

Von Meyenn, Ferdinand, et al. "Forget the Parents: Epigenetic Reprogramming in Human Germ Cells." *Cell* 161 (2015) 1248–51.

Vukojevic, Vanja, et al. "Epigenetic Modification of the Glucocorticoid Receptor Gene is Linked to Traumatic Memory and Post-Traumatic Stress Disorder Risk in Genocide Survivors. *Journal of Neuroscience* 34 (2014) 10274–84.

Waddington, Conrad Hal. "Canalization of Development and Genetic Assimilation of Acquired Characters." *Nature* 183 (1959) 1654–55.

Walker, Claire-Dominique. "Maternal Touch and Feed as Critical Regulators of Behavioral and Stress Responses in the Offspring." *Developmental Psychobiology* 52 (2010) 638–50.

Waterland, Robert A., and Randy L. Jirtle. "Transposable Elements: Targets for Early Nutritional Effects on Epigenetic Gene Regulation." *Molecular and Cellular Biology* 23 (2003) 5293–300.

Waters, Sonia E. *Addiction and Pastoral Care*. Grand Rapids: Eerdmans, 2019.

Weaver, Ian C. G. "Epigenetic Programming by Maternal Behavior and Pharmacological Intervention. Nature versus Nurture: Let's Call the Whole Thing Off." *Epigenetics* 2 (2007) 22–28.

Weaver, Ian C. G., et al. "Epigenetic Programming by Maternal Behavior." *Nature Neuroscience* 7 (2004) 847–54.

Wetzel, James. *Augustine's City of God: A Critical Guide*. Cambridge: Cambridge University Press, 2012.

Wing, Luman R. "On Gene Editing." In *Taking Persons Seriously*, edited by Mihretu P. Guta and Scott B. Rae, 293–314. Eugene, OR: Pickwick, 2024.

Witkiewitz, Katie, et al. "Relapse Prevention for Alcohol and Drug Problems: That Was Zen, This Is Tao." *The American Psychologist* 59 (2004) 224–35.

Wright, N. T. *Simply Christian*. San Francisco: HarperSanFrancisco, 2006.

Yan, Matthew Shu-Ching, et al. "Epigenetics in the Vascular Endothelium." *Journal of Applied Physiology* 109 (2010) 916–26.

Yehuda, Rachel, et al. "Holocaust Exposure Induced Intergenerational Effects on *FKBP5* Methylation." *Biological Psychiatry* 80 (2016) 372–80.

———. "The Public Reception of Putative Epigenetic Mechanisms in the Transgenerational Effects of Trauma." *Environmental Epigenetics* 4 (2018) 1–7.

Yong, Amos. "How Does God Do What God Does? Pentecostal-Charismatic Perspectives on Divine Action in Dialogue with Modern Science." In *Science and the Spirit: A Pentecostal Engagement with the Sciences*, edited by Amos Yong and James K. A. Smith, 50–71. Bloomington: Indiana University Press, 2010.

---. "Sanctification, Science, and the Spirit: Salvaging Holiness in the Late Modern World." *Wesleyan Theological Journal* 47 (2012) 36–52.

---. *Science and the Spirit: A Pentecostal Engagement with the Sciences.* Bloomington, IN: Indiana University Press, 2010.

Zablocki, Richard, et al. "Spiritual Practices in Addiction Recovery: Integrating Evidence-Based Strategies with Faith." *Journal of Substance Abuse Treatment* 73 (2017) 14–19.

www.ingramcontent.com/pod-product-compliance
Lightning Source LLC
Chambersburg PA
CBHW051937160426
43198CB00013B/2184